Love and War: Human Nature i~~n~~ ...sive
analysis of nothing less than the prospects for the future of the human
race. The authors are not only deep thinkers, but they are also experienced observers of humans and other animals. The book is full of
compelling stories based on their own experiences and studies in as
disparate places as Borneo and Banks Island in the Canadian arctic.
Their argument concerning the urgency of changing our ways as myopic,
self-indulgent consumers is indisputable.

If we can learn to love other life on this planet, we might have a chance
to save it and ourselves.

— **Robert Bateman,** World famous wildlife artist,
naturalist and author

Love and War: Human Nature in Crisis is a priceless gift to humanity.
We know the ravages of war within the world and within ourselves.
This book is a reminder of love's universal and eternal embrace, offering hope—a great guide for cultivating the seeds of love and eliminat-
ing the weeds of war. Read this book, look in the mirror and allow
change in your life.

— **Phil Jordan,** internationally recognized psychic consultant

Love and War deepens our understanding and brings us one step closer
to that distant day when we will lay down our sword and shield and
deserve the name *Homo sapiens*—wise human beings.

— **Sam Keen,** Best-selling author of *Faces of the Enemy*

This book is an excellent and well-researched integration of biology and
psychology and of science and fascinating personal experiences. It will
inspire us, every reader, to analyze our personal behavior and strive to
be more compassionate persons if we desire to survive this century
and leave our progeny a sustainable planet.

— **Peter Kolassa,** Esquire, Director of Hoffman Institute / Canada

You will receive valuable insights into modern biology, practical psychology and original research from reading this book, as well as enjoy amusing anecdotes and nice poetry. Most of all, you will be touched by the enthusiasm, honesty, and internationalism of these writers as they argue for a new way of living to create a new world order. To readers in the developing world, Asia, Africa, Arabia, and Latin America, such voices from America and Canada are heartwarming and encouraging. A better world is possible.

— **Dr. Harbans Nagpal,** Psychoanalyst Psychiatrist
Paris and Delhi, India

This is a perceptive and insightful exploration of the two greatest features of being human, which will rule the future, for better or worse. To understand them is crucial to the long-term survival of our species. The academic content of this book is exceptional—accurate and well described for comprehension, but I love the personal stories! They add the perfect touch to inspire us to take the action needed to understand the crisis we have created and the critical need for us to change before another war destroys our world.

— **Jarmila Peck,** Renowned Paleontologist
Czech Republic and Canada

Love and War is a stunning tour de force which traces the darkest shadows of our progenitors on humankind's evolutionary journey through individual, societal, communal, and nation-state relationships. It is a compelling and wonderfully readable analysis of how innate latent noble and fearsome factors in the human psyche surface in varying circumstances. In demonstrating how the latter can collectively lead to hostilities to achieve tribal or national goals, this book provides profound psychological insights as to why nations in the twenty-first century still employ war as an instrument of national policy. The authors specify what we must do, as nations and individuals, to create a global community working in concert to deal with the now-recognized universal threats to life on Earth.

— **Harry J. Petrequin, Jr.** Retired U. S. Senior Foreign Service Officer
Former Faculty Member, National Defense University
Active member of Veterans for Peace

LOVE

— and —

WAR

Human Nature in Crisis

Rudolf Harmsen, Ph.D. & Paddy S. Welles, Ph.D.

Robert D. Reed Publishers   •   Bandon, OR

Robert D. Reed Publishers
P.O. Box 1992
Bandon, OR 97411
Phone: 541-347-9882; Fax: -9883
E-mail: 4bobreed@msn.com
Website: www.rdrpublishers.com

Editor: Marian Sandmaier
Cover Design by Cleone L. Reed
Front Cover Art: Rifle with Flowers, © 2010 Conor Sullivan
Art on page 128: Picasso's Guernica, 1937
© 2010 Estate of Pablo Picasso / Artists Rights Society (ARS), New York
Photo Credit: Art Resource, NY
Book Designer: Amy Cole

Mixed Sources
Product group from well-managed forests and other controlled sources
www.fsc.org Cert no. SW-COC-002283
© 1996 Forest Stewardship Council

FSC

ISBN 13: 978-1-934759-46-2
ISBN 10: 1-934759-46-5
Library of Congress Control Number: 2010924044

Manufactured, Typeset, and Printed in the United States of America

Dedication

To our grown children, their children
and members of their generations,
in whose hands we will leave our planet,
with the fervent hope that it will be
a sustainable and peaceful place
for them to enjoy their lives
and the natural world.

Evolution has been in charge of our survival since our beginnings. Religion has given us faith and hope that we will continue to survive. These phenomena have allowed us to love and to wage war. This book explores and explains why we will not survive without stretching beyond our reliance on evolution and religion in order to create a new route to a more peaceful future—on which our survival depends. The choice is ours.

Why do we keep busy
expressing our fear,
reinforcing our fear,
teaching and transferring our fear,
and preparing to murder those
we haven't yet learned to love?
Are we determined
to extinguish our entire species
with a series of random, casual
acts of ignorance and stupidity
powered by someone else's
ego and greed?
Can we ever learn to suppress
our own ego and greed?
We must.
~ Jerry Ackerman

Table of Contents

Preface

*I believe that which binds us together is stronger
than that which drives us apart.*
~ U. S. President Barack Obama

Love binds us together. War drives us apart. We are writing *Love and War: Human Nature in Crisis* as a prescription to educate our hearts and minds as to how we could end unjust war and create more enduring love. Through a careful exploration and integration of evolutionary biology, ecology, social history and contemporary psychology, we answer the critical questions of why it is so difficult to make love last and war end. Both love and war carry us to the edges of human experience. Both involve intense fear and passion. Both place our very souls on trial. One could save us and the other will destroy us.

Because evolution is geared to our survival through competition, it will not produce ethical and moral human beings who will survive our methods of mass destruction. Because religion allows us to depend on inexplicable forces beyond ourselves and also has been the cause of divisiveness, it has incited many wars and continuing violence. It is now critical that we use our higher levels of consciousness to become a more compassionate and cooperative species if we choose to survive this twenty-first Century. Where we end up will be determined by our willingness to give up the mythical thinking that we can continue as we have in the past, to accept that our planet is critically ill and that each of us is a member of a team, or family, occupying one planetary home, who must cooperate in changing our behaviors to heal the planet into sustainability.

In preparation for his last article on World War II, three weeks prior to V-E Day, Ernie Pyle, renowned war correspondent, wrote in 1945: "And so it is over. The catastrophe on one side of the world has run its course. The day that had so long seemed would never come has come at last."[1]

This was written in draft form just days before a Japanese bullet killed him. Since that time, there have been brutal wars in every corner of the world, which continue as we write. Since 1945, the term "on one side of the world," has little meaning as terrorism has caused disasters which keep us all vulnerable. The United States was awakened from our insouciance by the attacks of September 11, 2001, and the ensuing war in Iraq has not only spread terrorism, but has set us back several hundred years in our efforts to create a safer world. If there is a nuclear or biological war, we will all suffer, we will all pay. Many of us will die—or wish we had.

The only way we will be able to live in a safer world is to gain deeper insight into why we turn so easily to aggressive battles, including crimes of passion and destructive wars, to resolve conflicts and then be willing to make alternate choices based on cooperation and compassion. This book offers us a way to bid "farewell to arms," and a way to survive on a more peaceful planet.

References:

Q. Obama, Barack. *The Audacity of Hope: Thoughts on Reclaiming the American Dream.* New York: Random House Audio, 2006.

1. Tobin, James. *Ernie Pyle's War.* New York: Free Press, 1997.

Introduction:
The Embedded Crisis

*It seems clear that wars are not made by generations
and their special stupidities, but are made instead by something
ignorant in human minds and hearts.*
~ John Knowles

In the beginning of mankind's search to explain things through ancient mythology appeared the goddess, Aphrodite (or Astarte), who sprang from a combination of passion and aggression. These two strong emotional states are often paired and seem to have reigned together in love and in war since the beginning of recorded history. They each tap into our deepest fears and our highest hopes. Fear is the basis for war. Hope is the basis for love. Our greatest need is to be connected to others, to feel loved. Our greatest fear is that we might lose love, therefore evolution has hardwired our brains with these opposing tendencies: to seek love and to fight to protect those we love and what we love. Thus, love and fear, leading to wars, engage in endless conflict in every human psyche from the cradle to the grave.

Love and War: Human Nature in Crisis examines in depth the dynamics of how this happens and analyzes why it is so difficult for us to make love last and war end. As individuals, we must be willing to explore our internal conflicts, the duels we fight between love and fear. These conflicts unresolved lead us to deny our need for love and allow fear to rule our lives. When fear is in charge, we are in grave danger of extinction via the final war.

Wars, by definition, feed on fear, aggression and violence. Love also can involve passionate aggression, which sometimes gives way to acts of violence—euphemistically referred to as "crimes of passion." Why does extreme passion drive us to react in rash and irresponsible ways? Social and psychological research suggests that our perception of reality changes when we are agitated or aroused. Recent biological research confirms that emotions determine which parts of our brains are activated, and therefore how we are apt to behave when confronted with those stimuli that tend to arouse passion or fear. This is fairly obvious when we "fall in love" and see "the other" as the answer to all of our dreams and fantasies. It is equally obvious when we begin to perceive "the other" as "the enemy," or as evil, and begin to prepare for battle—as individuals, or as nations. Our perception of others can shift from friend to foe instantaneously when fear replaces rationality.

These perceptions are often in conflict due to our needs at any given time and our inherent opposing tendencies. When making decisions, we all engage in minor, and sometimes major, battles between our wants and our "shoulds," our emotions and our cognitions, our desire for personal fulfillment and our responsibilities to others, our need for freedom and for stability. These tendencies play out as conscious and unconscious duels, which cast shadows in our minds, likened to the shadows on the walls in Plato's "The Allegory of the Cave." The shadows in Plato's allegory represent illusions to which the people of the Republic cling because they are afraid to walk out into the light, which will expose their illusions and force them to deal with objective truth. They prefer to be prisoners to their illusions rather than deal with the complexities of their internal conflicts, which become the major cause of external battles.[1] We need to explore our opposing tendencies—as they lead us to a deeper understanding of human nature.

We often hold onto illusions to convince ourselves that we are accepted and loved, especially when we are afraid that we may not be loveable. We recognize that we may well be wolves in lambs' clothing, but we try to keep the wolf well hidden, which increases our fear that he will emerge and expose us. We create ways to identify ourselves as unique and different, while craving to be intimately connected. The truth is that we are each unique and that we are all intimately connected. We each have constant duels with inner demons and angels. We are all afraid.

This was powerfully illustrated by Plato's allegory over two thousand years ago and again by Anthony de Mello, a contemporary philosopher, in his poem, *The Truth Shop*:

> *I could hardly believe my eyes when I saw the name of the shop: The Truth Shop. The saleswoman was very polite. "What type of truth did I wish to purchase, partial or whole?"*
>
> *"The whole truth, of course." No deceptions for me, no defenses, no rationalizations. I wanted my truth unadulterated. She waved me on to the other side of the store. The salesman there pointed to the price tag. "The price is very high, sir," he said.*
>
> *"What is it?" I asked, determined to get the whole truth, no matter what the cost.*
>
> *"Your security, sir," he answered.*
>
> *I walked away with a heavy heart. I still need the safety of my unquestioned beliefs.*[2]

We believe it is time to examine some of our unquestioned beliefs and develop a deeper understanding of what might be our whole truth.

Following World War II, Buckminster Fuller declared that the day we dropped the first atomic bomb was the day humanity started taking its final exam. This book is offered as a survival guide to help us deal with the looming problems that could cause our sociological and ecological collapse, and to help us comprehend what must be understood to pass our "final exam." Prior to August 6, 1945, wars were fought with the assurance that the human race would survive, as only relatively small portions of our species were wiped out. However, when the first Atomic bomb was dropped and killed over 70,000 people in one minute—not to mention the over 100,000 that died within the next few years as a result of that bomb—we created the possibility of our extinction. Much has been written about the rationale for this act of mass killing, but little has been written that explains why reasonable men and women, especially when passionately aroused, turn so easily to fighting and killing to resolve problems.

In Tolstoy's classic, *War and Peace*, Count Rostoff, in a conversation with his soldier son, Nikolai, beseeches him to love passionately while he is at home on holiday, and then return to the war with the same amount of passion; love and war, he counsels,offer the only true bliss in life.[3] The only thing written about peace in Tolstoy's tome is that it is boring. Is this what many of us actually believe, although we profess to desire a peaceful world? How much peace can we tolerate without becoming bored? Must these questions remain mysteries, or is it time to search seriously for some answers? In our search for answers, we have explored the biological, social, and psychological literature, which will be summarized throughout the text. We have also studied the social and ecological conditions of our pre-sapient ancestors and close relatives and believe their lives hold some relevant clues.

Freud espoused the contention that war is inevitable given the selfish, power-seeking nature of human beings. However, he failed to address the possible reasons behind our warlike nature. History might well support Freud's pessimistic view, as there have been only twenty-nine years in the recorded history of humankind during which a major war was not being fought. This does not even begin to touch the less than major conflicts between tribes, families, siblings, other personal rivals, and even lovers, who often wage vicious battles, sometimes ending in death. Relatively sane people who claim to hate war, often will say, "I could kill you," when engaged in an argument with a lover or close friend. Is aggression to the point of killing another of our species for material gain, or to prove that we are right, an innate aspect of our nature? We shall see.

Definition of Terms

Love is perhaps the most over-used and least understood concept in most spoken languages. We are aware that there are different kinds of love, various levels of love and special moments of love. All love does not have an erotic component; however, the dominant definition in literature and in this book is an intense emotion of passionate, erotic attachment to another. This type of love is the most apt to cause us to "go nuts," to behave irrationally—even to kill. It also allows us to feel more vibrantly alive, to feel beautiful and to want to live forever. It can contain joy, desire, fear, jealousy, desperation and violence. We seek it and we hide from it; rarely do we trust it. This deep desire for what we fear we may not be able to obtain or sustain is the major raging battle in the human psyche.

War is conflict—win or lose, or carried to the extreme, live or die. It can happen within the context of love relationships. Most wars begin within ourselves as conflicts between various needs or desires and our fears, between what we perceive to be right or wrong, good or evil, and what we want to do vs. what we feel we should do. Wars spread from internal to external when we begin to project our internal conflicts onto others, which began in childhood within our families. Our call to arms against others is usually based upon fear, pride, or a perception that the other is bad—or even evil. Wars between races and nations are extensions of the above phenomena. They often happen due to a kind of national malignant narcissism and generally occur when more appropriate resolutions are abandoned.

Passion is the experience of emotional intensity. The original translation in Aramaic is suffering, and there often is an element of suffering within a passionate experience. Photographic research of animal and human faces during orgasm show that it is impossible to distinguish pleasure from pain. The obvious emotional experience that can be observed (from the neck up) is intensity. In this book, we define passion as an intense, overwhelming, driving emotional state, which can defy reason and rationality.

Aggression is psychobiological energy that initiates verbal or physical confrontation with another, or group of others. Aggression can be used to influence, to protect, or to defend oneself, others, or property. We believe that aggression can be used to solve conflicts in both a positive and negative way, although it has received more negative than positive press. It is a useful trait in most competitions, but can become harmful if carried to an extreme. It can be observed in healthy play and even in social reconciliation.

Natural selection is the proper term for what is colloquially called "survival of the fittest." It describes an essential aspect of the interaction between individual living organisms and their total environment. The established fact that among all individuals of a species no two are exactly similar means that each individual organism interacts with its environment in a different manner. This implies that some individuals will interact with their environment in a manner which results in their surviving and reproducing more successfully than others. This process of differential survival and reproduction is called natural selection. Charles Darwin was the first to recognize the immense importance of this simple process. He realized that any differences among individual organisms of a species that were genetically determined would be affected by natural selection. Some of these genetically determined traits would survive better than others, resulting in small changes in the average traits of a species from generation to generation. Over long periods of time, this can lead to major evolutionary changes, provided that there is also a mechanism to introduce new genetically determined variation into living organisms. There is such a mechanism in random mutations in an organism's DNA.

Many interpret "survival of the fittest" as the survival of the strongest and most aggressive through evolution, but that is not necessarily the case. Whether strength and aggression or cooperation and meekness will be selected depends on the ecological and social conditions at the time the selection takes place. All traits come with both a cost and benefits; natural selection acts on all traits simultaneously.[4]

Religion is a general term applied to belief systems that are based on the existence of deities with supernatural powers that transcend human power. Most religions began as myths to explain the wonders and mysteries of life that had not yet been scientifically explored or understood. Because we humans seem to possess a sense of the numinous, a spiritual quality that inspires us to go beyond a literal explanation of natural phenomena, to move from scientific into sacred, we created gods and goddesses and ascribed to them divine powers. The earliest ideas of gods and goddesses seem to have surfaced along with cave drawings, the earliest art.

Purpose

The major purpose of our book is to get us in touch with the ways we employ aggression and passion to meet our personal needs and to achieve our ambitions, goals, and dreams. Experiencing our aliveness is our deepest psychological need from the cradle to the grave. Feeling passionately alive is the ultimate gratification of that need. Must we behave irrationally or violently to attain that gratification? When does love become irrational? Is war ever rational?

This book explores the importance of rationality and the need to control irrationality in organized society. To what degree can an individual or society remain stable under threat, instead of becoming irrationally violently aggressive? We approach this question through the study of social animal groups, including *Homo sapiens* and its recent ancestors. We are aware that culture, *per se*, is not biologically transmitted. Yet, there is biological research indicating that there are genes that cause some of us to seek higher states of arousal, to react more irrationally and with more force when we perceive ourselves to be physically, territorially, or emotionally threatened. Is it justified to label that which threatens us as evil, while we perceive ourselves as good, and then to think that destroying what we perceive as evil will remove evil from the planet? The dividing line between good and evil is located in each of our minds.

All relationships and social groups must maintain an interdependence to exist, while as individuals we struggle with creating a balance between the personal needs of our egos and our social identity. The need for social identity requires an acceptance, or at least a tolerance, of differences; yet differences seem to trigger our fears. Is it the fear that others, different from us, may overpower us or gain control of limited resources that causes us to wage war?

In the following chapters, we examine the essential questions about why we are the way we are, why natural selection has burdened us with very complex personalities, and how we must adapt rationally to a world filled with irrational acts. This becomes the most difficult challenge of man's responsibility to survive. We propose that it is time to make different choices that will implement more peaceful ways to solve the crises we are now facing. We must create an exciting, passionate dance between our individual selves and others, regardless of differences.

We do not claim to have learned every step of the dance, but as careful observers of our own emotions and behaviors, which are not different from yours, we are committed to understanding the mysteries of conflicts so that we may all dance together. We believe that understanding our evolution and living from our higher consciousness will help us come to terms with human nature and control those aspects that are inappropriate for life in a civilized world.

The Dance
When the snow falls, the flakes spin upon the long axis
that connects them most intimately
two and two to make a dance.
The mind dances within itself, taking you by the hand,
your lover follows
there are always two, yourself and the other,
the point of your shoe setting the pace,
if you break and run, the dance is over.
But only the dance is sure.
Make it your own.

Who can tell what is to come of it?
In the woods of your own nature
whatever twig interposes,
and bare twigs have an actuality of their own
in this flurry of the storm that holds us,
plays with us and discards us
dancing, dancing as may be credible.
~ William Carlos Williams[5]

References:

Q. Knowles, John. Great Britain: Secker and Warburg. 1956.

1. Klein, T. B. Edwards, T. Wymer. *Searching for Great Ideas.* New York: Harcourt, Inc., 1998.

2. De Mello, Anthony. *Awareness: The Perils and Opportunities of Reality.* New York: Doubleday, 1990.

3. Tolstoy, Leo. *War and Peace.* Translated by Alexandra Kropotkin. Great Britain: The John C. Winston Company, 1949.

4. Shermer, Michael. *Why Darwin Matters.* New York: Holt McDougal, 2007.

5. MacGowan, Christopher. *The Collected Works of William Carlos Williams.* New York: New Directions Publishing Co., 1986.

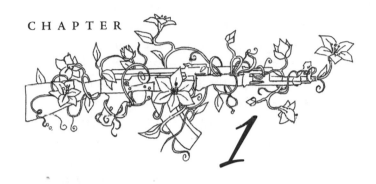

1

From Carbon to Conscience

What a piece of work is a man!
~ Shakespeare, Hamlet

An elderly Cherokee was teaching his grandchildren about life. He said, "A fight is going on inside me, a terrible fight between two wolves. One wolf is evil; he is filled with fear, anger, envy, greed, arrogance, guilt, blame, shame, lies, self-involvement and vanity. The other wolf is a good wolf. He is filled with love, hope, compassion, humility, generosity and faith. You have the same two wolves fighting inside of you and so do all other people." The children thought for a moment and then one child turned to the grandfather and asked, "Which wolf will win?"

The old Cherokee replied, "The one you feed."

We all question the reasons for the endless conflict between these two internal wolves. Each of us longs for the extended contentment we imagine would fill us if the bad wolf would just roll over and give up.

But, he's a tough, old—very old—aspect of our minds, as is the good wolf. Actually, we need for them to co-exist. They engage in a constant dialogue, one that helps us choose between what we want to do and what we feel others want us to do, between our heads and our hearts, between what we have been taught or intuitively feel is good or evil.

Since the beginning of time, we have been told stories of good and evil, and yet, there is neither good nor evil to be found beyond human-kind with its social and cultural constructs. Could it be through our judgment of a situation, our perception of how we might be affected by it, that we label something good or evil? Each of us holds the potential to enhance life or to cause harm, to create love or war—battles within inti-mate relationships included. Every action has consequences, and some-times an action will generate both a positive and a negative consequence. The consequence of an action is often contingent upon our perceptions, expectations, attitudes, points of view, or a situational circumstance, such as the difference between winning and losing a competition. The winner will view the situation as positive, whereas the loser will usually view the same situation as negative.

The question we must address is: Are these opposing forces of good and evil really a part of human nature or are we dealing with traits devel-oped over time through our specific cultural environment? To answer this question may be of immense importance for our future survival. We must go back a long way to what is certainly the most important revo-lution in the history of our Earth, which happened roughly 3.5 billion years ago, when the first life forms arose on the previously lifeless Earth. Nobody knows exactly what happened. All we know is that somewhere on this planet a very small lump of non-living organic matter came to life, that is, began to interact with the environment in such a way that an energy exchange was established, setting off an entirely new direction for the world's future. Although we do not know what the first life form looked like, or how it came about, we can say with fair certainty that the Earth changed drastically very soon after the origin of life.

What Life Is

We know that all life, of which we are aware, is essentially carbon based and was made possible through the unique nature of planet Earth. It had a chemical composition rich in carbon and its distance from the sun provided a temperature range that allowed complex, organic matter to form and survive. Exobiologist, Stanley Miller, illustrated experimentally in 1953 that a combination of simple compounds: methane, carbon and ammonia, all present on Earth prior to life, when combined with energy (electricity) will form more complex molecules, producing amino acids, the building blocks of life. Recent theories, supported by fossil evidence point to volcanism, deep sea vents and four billion year old meteorite showers as alternative possible sources for the formation of larger organic molecules.

Life is the property of a localized quantity of matter and energy, which allows that entity to consume high-quality energy from the environment (food) and to release lower-quality energy (waste). This may sound like a vague definition, but it isn't. It is based on one of the world's most important laws of physics: the Second Law of Thermodynamics. This law states that the energy contained in a closed system inevitably and spontaneously degrades by acquiring entropy over time. Entropy is simply the measure of how far a unit of energy has degraded toward being unable to do any more work. If a living organism were a closed system, its energy would slowly degrade, making it impossible for the organism to perform the work necessary to maintain itself. In short, it would die. When a system over time acquires entropy, it will show a loss of structure, organization, direction and information.

An example might be your grandparents' wind-up clock. When your grandfather winds up the clock and sets the time, he introduces mechanical energy (low in entropy) into the clock mechanism. This energy then runs the clock and in doing so, it gains entropy. Not one unit of energy is lost, but it acquires entropy as it changes from concentrated mechanical energy to heat and sound, which is no longer capable of keeping the clock going. This process, a declining quality of energy, would be the result in any system, unless more low-entropy energy is gained from an outside source. Since all organisms constantly interact with the environment, they are considered open systems. A living open system avoids being degraded by adding new low-entropy energy to itself and by discarding useless high-entropy energy. The process that controls and directs this interaction is life itself—the most amazing and extraordinary process that exists in this world.

As a corollary of the self-maintenance property of life, we assume that growth and reproduction must have followed soon, because without it, non-reproducing life forms would have sooner or later met with accidental death, and life would have ceased. What is important is that there are only two major sources of high-quality energy available on Earth. One is sunlight, which "feeds" plants. The second is other organic matter, which is the food of most animals. The moment we realize this, we see in stark outline the three enduring dictums of animal life: eat others, avoid being eaten by others, and eat before another grabs your meal. Ecologists would phrase it as predation, predator avoidance and competition. This holds for bacteria, elephants, apes and humans. Yes, Dr. Watson, nature is not a bed of roses.

Survival of the Fittest

Let's take a closer look at how life drastically changed the world. Three processes started to happen. The early life forms started to consume any high-quality non-living organic matter, which was a part of the early Earth. Next, they began to devour one another. If this were all, the Earth by now would be no more than a planet populated by a mass of very primitive, inefficient life forms. But here we introduce another important scientific law, the Law of Organic Evolution. This law of nature will play a leading role throughout this book. Despite its basic simplicity, this law is not well understood. It is best to approach it as a two-stage logical argument. First, the law states that when living organisms grow and reproduce, they will become more diverse, simply because, like all complex processes, growth and reproduction are not perfect. This leads to populations of individual organisms that inevitably differ in structural and behavioral traits from one another. Despite the old adage, two peas in a pod are *not* identical. These differences in structure and behavior imply differences in abilities to compete and survive in the world's highly competitive environment.

The second stage of the law states that in a world of limited resources, some individuals will be better adapted to prevailing condi-

18

tions than others. This, in turn, results in varying abilities to survive and reproduce, which is properly referred to as "natural selection." When we combine the two parts of the law, it becomes: *Species of living organisms evolve over time due to natural selection acting on randomly occurring variation.* The biological processes involved in evolution, such as genetics, development, adaptation and sexual reproduction, are fiendishly complicated, but they need not be understood to grasp the basic principle. Professional biologists have studied these complex processes in much detail, and it is encouraging that despite a great deal of diversity in how evolution occurs in different groups of organisms, the general principle, as stated above, is always the same.

During our three and a half billion years of evolution, life forms have changed dramatically. There is no reason to suggest that the process of change will not continue until the end of time. Different organisms have developed many new ways of surviving in our "dog-eat-dog" world. Yet, contemporary scientists agree that we all share a single, unique origin of life—*all forms of life have evolved from the first living entity.* If this strikes you as overwhelming, it is. In the most complex social life forms, such as wolves, whales, apes and people, intricate methods of survival have evolved. Some of these depend on deadly competition; others depend on cooperation, trust and altruism. Many animals have advantages over humans when it comes to endurance, speed, strength, instinct and mastery in water and air, but humans seem to have mastery of the ability to think things through—to reflect, to imagine beyond what we know and to consciously create change. This ability makes us who we are. It is our blessing and our curse.[1]

The Evolution of Consciousness

Consciousness can be seen as the initial stage of what we call intelligence, the faculty that gives us the capacity to make choices—including which wolf should be fed in any given situation. Our choices and the behaviors that follow express our humanity and our inhumanity. Only during the past fifty years has it been scientifically acceptable to address the issue of the consciousness of animals. However, most biologists and psychologists agree that an animal has consciousness when there is empirical evidence of its ability to emote moods, to evaluate a situation from its personal past experience and to implement an appropriate course of action that goes beyond native instinct. An example is the formation of political alliances among members of a group of chimpanzees to counter the dominance of a despotic alpha male. Philosopher of science, Karl Popper, in *The Self and Its Brain*, states: "The emergence of consciousness in the animal kingdom is perhaps as great a mystery as the origin of human life. Nevertheless, one has to assume, despite the impenetrable difficulty, that it is a product of evolution, of natural selection."[2]

Trying to make sense out of the evolution of consciousness becomes less of an obstacle when we realize that consciousness did not appear suddenly; we did not wake up from a long primordial sleep into full consciousness. Instead, several levels of consciousness developed slowly over evolutionary time. A general awareness of the environment and the ability to respond in an appropriate manner is a basic form of consciousness, while the ability to assess one's place in a larger time frame from before birth to after death requires a more complex, sophisticated level of consciousness. When considering the origins of our awareness, it may help to unravel the disjointed thoughts that pop into our minds to examine a less sophisticated version of animal consciousness.

The Life of the Meadow Vole

When a meadow vole runs along a path it has laid out through a field, we can interpret the small rodent's movement and progress along the path in different ways. A simple mechanistic interpretation could be as follows. The vole has left a scent trail every time it runs along the path, and each new time on the path, its olfactory receptors pick up the lingering scent. A set of messages from its nose is processed in its brain, and an innate response makes its musculature propel the vole along the path. The vole is not conscious of what it is doing; it is a complex organic robot. Only a stronger, overriding sensory input, such as the smell of a predator would cause the vole's brain to alter its activity pattern. Its life in the field is only a slightly more complex version of the life of an earthworm or a dandelion. It is responding to information gathered by its sensory receptors. This information is combined in the brain with coded and stored information of its past life-experiences in the field and processed according to a genetically determined set of rules. Its brain then sends messages to all body parts involved in producing the necessary responses to maximize the vole's survival.

Such a strictly mechanistic interpretation of the life of a fairly advanced life form is problematic for a variety of reasons. It is especially hard to imagine how such a simplistic version of an animal's life can serve the immensely complex set of responses needed for an animal to survive. A more sophisticated interpretation is needed. Few people would seriously argue that the vole, using vole language, consciously and rationally designs its responses to its environment, presenting to itself such thoughts as (English translation): "The last time I went to that eastern edge of the field looking for clover, I smelled a weasel, so perhaps today I will go to another part of the field." However, we must come up with a version of the meadow vole's life that takes into account the complexity of the vole's world and accounts for its necessary and intricate responses.

Voles live in a highly complex world, with a three-dimensional architecture consisting of many different materials, most of them either alive or of organic origin. They also live in a social environment, based on complex relationships with breeding partners, litters of juveniles and other more or less related individuals. The individual vole faces a constant battle for survival, having to respond to competition, predation, food shortages, and harsh physical conditions.

Like all animals, voles have a genetically determined set of capabilities to respond to their environment. However, their genes must provide them with much more than a simple input-output system. It must include the ability to learn new responses and to improve the skill with which the responses are executed. When the vole moves along its path, it calls on the coded information stored in its brain, but it does not respond like a mechanical robot. The vole conjures up a sequential set of mind pictures depicting the path up ahead. These are not like the pictures a small TV camera would see; they are pictures that combine olfactory, auditory, tactile and visual elements. The pictures also have a certain level of possible variability, as time of day, weather, and random events are within the realm of possibility. At every step, the reality of the vole's immediate surroundings forms a new short sequence of pictures, which its brain compares with all possible expected pictures.

When the vole goes off trail, venturing into new terrain, the possible breadth of expected pictures becomes even wider, but fuzzier at the same time. The vole now has to generalize and synthesize composite pictures, some safe, others dangerous, some useful, and others wasteful. At every step, the vole has to compare the real pictures provided by its senses with a mass of remembered and/or synthesized pictures. Then it has to decide how to respond so as to satisfy its hunger and other drives, keep warm and dry, act appropriately in its social environment, and avoid predators. The question we want to pose at this stage is the following: Does the life of the vole as described above imply that the vole is conscious of itself and its role in the world?

We believe that the vole probably is incapable of forming anything other than self-centered pictures of its real and possible environments. It cannot see itself in the field the way a vole-hunting kestrel would see it from above, or how a potential mating partner would see and evaluate it. It does not have a formed concept of how long it will live or of what death implies. However, when it is hungry and on a foraging expedition, quibbling with another vole over access to a choice bit of vegetation, and the scent of a fox suddenly hits its nose, it has to design quickly a complex response. The vole shows convincing evidence of intent, the capability to synthesize appropriate responses, unique to the situation at hand and will experience an emotional state that we would recognize as panic and fear. If our vole managed to survive this encounter with a fox, he will learn from the experience and have an improved chance of surviving the next encounter as well. Does this mean that the vole is conscious? The answer depends on our definition of consciousness and on how we humans choose to respond to our environment.

A Biologist in the Meadow

My wife and I (Dolf) live in the country and have a hiking trail, which loops around through a wooded area, but it begins through a hayfield with a vole population. I recognize the mini-trails that these small rodents make through the grass in summer and through the snow in winter. Every now and then, I see one of these small, dark short-tailed rodents. They are astute, can move remarkably fast on short legs and can disappear so suddenly that it feels as if they have simply changed into a bit of grass. But despite their highly efficient adaptation to their environment, they are often the victims of predation; their lives are short, for which they compensate with large and frequent litters. Before writing the above section on the life of the vole, I went for a walk on our trail and tried to figure out how I make decisions about how to live my life moment by moment. I usually use language when I think about what I am doing, but we humans do not have to use language to enjoy a variety of experiences.

On my walk through the field, I chose to avoid language. I focused directly on what my senses perceived in my environment, based on past experiences in the field. I merely created pictures of how it looked and smelled, how the wind touched my face and how the sun sparkled. I pulled older pictures of previous hikes out of my memory banks. I felt at ease, because what actually surrounded me fit smoothly and comfortably within the realm of my mental picture book. Only because I had planned to investigate my walk through the field was I conscious of my usually subconscious constant comparisons of what I walked through with what should and could be encountered. It became clear to me that on a normal walk, I do not consciously make frequent decisions about every step I take; at the same time, however, I am conscious of my environment. The vole, I believe, is also conscious in that sense of the word.

There are, of course, differences between the way the vole sees its environment and my way. I am here, writing about the vole and myself, because I am consciously thinking about our similarities and differences. I doubt the vole thinks like this. Thinking in pictures is hard to extend into philosophy. For that we need a more sophisticated language, which only the human species has developed. Voles, elephants and chimpanzees think in pictures—complex, moving pictures. Picture language has the grammar of the environment: things are nouns, actions are verbs, differences can be seen as adjectives and adverbs, etc. Concepts such as love, sadness and other emotions add feeling to the mind-pictures. Human consciousness has a communicable complexity to it, which is lacking in most other animals, but we would make a big mistake if we assumed that all other animals have no consciousness.

Animal I.Q.'s

Consciousness implies that there is a high level of evolved complexity to the brain. Although we do not yet have I.Q. tests for animals, we can often observe consciousness when animals play. While on the Lindblad expedition-ship *Sea Lion*, observing young sea otters at play along the boulder strewn shore of Idaho Inlet in southeastern Alaska; a group of us were amazed by how similar their games were to the games we played as children. They played hide and seek, tag, and king of the boulder, while their mothers looked on, and even disciplined when necessary. It was easy to imagine which otters seemed to feel excluded, which were considerate of others, and which would strategize in game playing. After an hour of observation, we realized that we were speaking about them as if they were children. Several passengers began to relate stories of how their domesticated animals and pets played games and exhibited emotions, especially in showing devoted attachment. We questioned if this could be the same emotion as love between human persons.

Sea otters are highly social animals and exhibit signs of having developed a cultural aspect to their social behavior. They spend a lot of time just "hanging out" in groups called rafts, because these groups often consist of animals just floating on their backs somewhere in the middle of a bay or inlet. Within such rafts, the individuals interact extensively; they swim from one to another, engage in mutual grooming, and new arrivals are greeted as they swim into the raft. Sea otters also use tools, but there is no evidence that they make or re-shape tools. At that level of social structure, one would expect the animals to need a fairly sophisticated level of consciousness to maximize the advantages of the network of social interactions.

By far the most advanced social structure in the animal world is found in what we know for sure are our closest relatives, the chimpanzee and the bonobo. These animals provide us with the best evidence for consciousness. These apes have a far more developed sense of self, in that they recognize themselves in a mirror, which even monkeys don't seem to be able to do. They also communicate with one another in much more sophisticated ways. Christopher Boehm in *Hierarchy in the Forest* describes in detail how these animals form political alliances that can determine which individual will play the alpha role in a troop. These species also have strict socio-cultural rules, which regulate such important activities as food sharing and mating.[3] It is obvious that these animals have evolved a high level of consciousness.

Recent research strongly indicates that not only chimpanzees are able to associate words in human language with the objects the words represent, but also dogs and dolphins are capable of making such associations. These animals have demonstrated that they can select a specific toy from a group of toys when asked to do so. They are also responsive to verbal commands by humans, demonstrating that they have, to some degree, gained an understanding of our verbal communication system. We humans are possibly in for many surprises when we make the assumption that we are the most advanced species—perhaps in the ways we tend to measure intelligence, but perhaps not when it comes to using different modalities to communicate.

Reason and Intent

Neurobiological research indicates that ancient, pre-sapient, parts of our brains allow us to experience emotion, and that emotions cause us to behave in certain ways, including to make love or to make war. However, it was not until the development of the neo-cortex that we could evaluate the consequences of that behavior. We know that about one hundred million years ago several new layers of brain cells were added to the brains of mammals to form the neo-cortex, described as our "seat of thought." And we know that the *Homo sapiens* neo-cortex is considerably larger than that of other mammals, which should give us an intellectual edge, as it allows us to think about our emotions and judge their effect, rather than only react impulsively to them. Like the human biologist in the meadow, we may choose to omit language in order to have a purely sensual experience. But to contemplate that experience, we utilize thought and reason, which add a more extensive dimension of consciousness to our existence—and to our relationships.

Psychologist Daniel Siegel, in *The Developing Mind*, states that the mind emerges from patterns in the flow of information within the brain and between brains. The mind is literally created within the inter-action of internal neuro-psychological processes and interpersonal (or "inter-animal") experiences, which shape the maturation of the nervous system. This simply means that our interactions with others, or the lack of interactions, affect how our brains develop.[4] Most psychologists agree that it is within this *developing via relationships* mind that we manifest who we really are. Each interactive experience from the moment of our birth leaves an imprint in our minds, i.e., makes a mark that is stored as a memory, either conscious or unconscious, but that can be remembered when stimulated by a similar experience. This is easily observed when an infant responds to a loving caretaker, whose presence is associated with having the baby's needs satisfied. The baby feels content and safe when the caretaker comes close. When a stranger, such as a physician with an instrument in her hand, comes close to the infant, who has felt discomfort or pain in a past experience with such an entity, there will be obvious distress, triggered by fear in the baby's mind. This same phenomenon is frequently observed in many animal species. Watch a group of squirrels busily gathering acorns when a large bird flies overhead. The shadow of the bird is enough to bring all activity to a halt. The squirrels freeze in fear. We do the same when the shadow of what we fear looms large in our minds.

The difference in our level of consciousness and that of the squirrels is our ability to think and communicate about our emotions and to design a strategy, based on reason and intent, to cope with our reactions. Natural selection can only select traits that are adaptations to the past with short-term fitness advantages. It is blind to the future, and hence cannot design traits for long-term survival. However, natural selection has resulted in our having the power to reason and to be able to intentionally plan for the future. The power of reason has also allowed us to formulate social concepts, such as ethics, morality and social responsibility. The human mind does not merely consist of a spectrum of emotional states from violent aggressiveness to loving compassion. Our minds give us the ability to manage our emotions with degrees of rationality and to make decisions that are not just impulsive reactions to emotional states.

Apes or Angels?

Disraeli questioned whether man is ape or angel, or a bit of both. Serious thinkers and observers of human behavior continue to argue over the origins of good and evil in our lives, but most agree that the concepts grew out of our need to judge and try to control the behavior of others. What has never been agreed upon is whether some of us are inherently "good" people and others inherently "bad" people. Do we *come* that way, or do we *become* that way? It would be fairly easy to obtain a safe world if there were only some people who would commit evil deeds. They could be designated and separated out from those of us who would only commit good deeds, but nature presents us with a more troublesome truth: *the dividing line between good and evil lies deep within the mind and heart of each one of us.*

Ask yourself: Why do I have conflicting emotions of love and tenderness as well as anger and hate towards the same person? How can I feel love for my spouse or best friend in one moment and be furious at him or her the next? Why am I attracted to some people, while feeling ambivalent or strongly negative toward others? Then ask yourself where each of those feelings originated. These conflicts are partly triggered by past experiences that left their positive or negative responses in our memories. Our parents both cared for us and wounded us—often in the name of love. When we become aware of such contradictions, we begin to categorize things as good or not good. Religious beliefs are often formed through this same type of categorization.

But our experiences are only a part of the story. Prior to our having experiences that leave us in conflict, we are born with a genetic predisposition that influences us to react and behave in various ways. We each have many thousands of interacting genes that affect one another in subtle ways to form our unique individual genotype at conception, which immediately is influenced by our environment to produce the persons we become. Some genotypes more easily develop into caring, tenderhearted individuals; others have to work hard at becoming such a person—and might not be happy playing that role. Some are born with the potential to become tough competitors, while others prefer to remain passive and compete in a different way. Often these more passive souls appear weak, but tend to learn to use weakness as a way to manipulate the more aggressive. Whichever is our natural tendency, the counterpart lies close by, and under stress, can leap into action. One of our major challenges is determining when and to what extent we should behave aggressively, as opposed to when it might be to our advantage to "just let go." Behaving from either extreme end of this passive-aggressive continuum for extended periods is usually destructive, for we set ourselves up to become either warmongers or doormats.

Historically, much more attention has been focused on our aggressive, more selfish traits than on the more altruistic, unselfish aspects of our nature. Where does our caring, loving side come from? Freud wrote that contradictions seem to lie peacefully next to each other within the human mind, until a specific set of circumstances forces a choice. We may believe that we are fully committed to the peaceful wolf in our persona and have successfully suppressed the war wolf, but we should not be too smug. Certain social conditions and crises have shown that when we are stressed or manipulated beyond a certain point, most of us tend to display our war-like wolf—and often, rightfully so. In many cases, our aggressive wolf can be a saving ally.

Humane Apes

The most widely held interpretation of human nature is that we are born carrying the selfish, vicious wolf in our hearts, and all goodness must be acquired by learning and absorption from the civilized world. This Hobbesian theory is referred to as the "veneer" theory by Frans de Waal, a contemporary zoologist and ethnologist specializing in the study of primate behavior. He points out that our close relatives, the chimpanzees and bonobos, reveal a more complex and realistic origin of human behavior. His research presents a compelling argument for looking to our animal ancestors for possible roots of more noble behaviors. In his book, *Good-natured: The Origins of Right and Wrong in Humans and Other Animals*, he reports the following story of Mai, a high-ranking female chimpanzee, and her fellow chimps at the Yerkes Primate Center, in Atlanta.[5]

One afternoon as the entire colony gathered around Mai, de Waal observed a high level of emotional expression throughout the chimpanzee colony. All of the apes were silent and staring at Mai's behind. Mai stood half upright, holding one hand between her legs, while another older female mimicked her by cupping her hand between her own legs. After a brief time, Mai tensed, squatted more deeply and gave birth to a baby, catching it in her cupped hands. Mai's best friend in the colony, another female named Atlanta, let out an excited scream and began embracing the chimps closest to her.

During the following weeks, Atlanta was carefully attentive to Mai's needs, defending her, grooming her, and frequently observed staring at and gently touching Mai's new son. It seemed to de Waal and other observers that Atlanta was expressing empathy for her friend, Mai, as well as expressing love towards the new baby. Empathy is considered the second stage, following attachment, in the development of a human's ability to love. It is also the cornerstone of human morality.

Interestingly, another female chimpanzee in the compound, Georgia, was as selfish as Atlanta was altruistic. Georgia never shared, behaved aggressively towards Mai the day after the birth, and was avoided by the other members of the group. She rarely shared with her two daughters, who would roll on the ground screaming for food and attention, but Georgia continued hoarding food and fresh branches for herself, while her daughters suffered. Is this difference in behavior due to genetic make-up, or is it a reflection of Georgia's childhood, or periods of suffering later in her life? Could it be a complex combination of both?

Altruism

An interesting twist was woven into the arguments over the origins and meaning of good and evil by Sigmund Freud, who believed that all behavior stemmed from selfish motives. From his point of view, Atlanta's behavior towards Mai may not really be as unselfish, or self-sacrificing, as it appears. Freud believed that when we choose to behave altruistically, we do so in order to accumulate social benefits, such as another person's gratitude or friendship, which are essentially self-serving.[6] However, Sigmund Freud lived in the nineteenth and early twentieth centuries, well before the first scientific primatological research started to reveal the complex social biology of our nearest animal relatives. We now know that the seeds for the evolution of ethical, altruistic behavior were sown long ago in our distant ancestors. Yet the evolution of true, unselfish behavior remains an enigma. How can self-sacrifice to save others be selected for in nature? We also know that typical natural selection, which selects individuals who have traits that increase survival and reproduction, is not the only evolutionary force in nature. There are two other modes of selection: kin selection and group selection, interacting in complex ways that affect a population over time and could be responsible for the evolution of altruism.

The kin selection argument simply states that when an individual expresses a trait, which is costly to the individual, but beneficial to relatives, a kind of evolutionary cost-benefit analysis takes place. If the cost (reduction in fitness to the individual) is less than the benefit (increased fitness of relatives with the same trait), natural selection will result in the spread of that trait. For example, if a teenager who is a strong swimmer sees her younger brother fall into a fast-running river and jumps in to save him, she is only taking a relatively minor risk. She is subconsciously confident of this, and hence, of the probable survival of some of their shared genes, whereas letting her brother drown, would reduce the presence of their shared genes in the population. Up to this point, it could be claimed that the altruism is still suspect. It may appear heroic for a sister to risk her life to save her brother, but Freud could still argue that there is a benefit for the sister, "hidden" in the brother's survival. But, how do we explain an altruistic, self-sacrificing action to save a non-relative?

The answer to the question of how true, unselfish altruism evolved in our ancestors has been described as originating through the interplay of reasoning power with a social structure based on small, competing, nomadic groups. For at least a couple of million years, our ancestors lived in small groups, which competed and fought with other such groups. This constant violently agonistic behavior often resulted in some groups surviving and spawning new groups, while others fell apart and died off. It is not hard to imagine that groups with a high proportion of cooperative, self-sacrificing individuals would be more likely to survive and thrive than groups with chronic leadership problems and other forms of internal strife. Nonetheless, most evolutionary biologists consider group selection to be too weak to overcome individual selection favoring selfish behavior. However, a combination of kin-and-group-selection could have brought our ancestors to a point where at least among relatives within groups, altruistic behavior became recognized and valued. At this stage, reasoning individuals would start to form rules pertaining to within-group interpersonal relations. This has led to the formation of yet another form of natural selection, *socially driven selection*, which will be discussed in Chapter Four. We continue to be works in process.

Works in Process

In fairly primitive animals, such as fish, several species show remarkable tenderness and the ability to care when it comes to dealing with their own offspring. Male sticklebacks, which are small predatory fishes, carefully guard the swarm of their newly hatched offspring. When an overly adventuresome baby fish strays from the swarm, the father will gently take it into his mouth and spit it out back into the swarm, where he can keep a close eye on it. Similarly, large mother crocodiles also carry their much smaller babies in their heavily toothed mouths. Such behavior is probably completely innate and not consciously executed, but it illustrates how an attitude and action of tender care can evolve as a behavior. At the same time that sticklebacks are protecting their own babies, they may be observed devouring the babies of other males of their own species. Wolves and apes are also known to harass, fight and kill others of their species, even others of their own group. When we study the social behavior of apes, such as chimpanzees, it becomes clear that aggressive, confrontational behavior and cooperative, kind behavior are integral parts of their personalities, both essential for the maintenance of their social structure.

What we see in these apes, we must see in ourselves—not two separately evolved personalities, one evil and one good, but one integrated personality with widely different, even conflicting, aspects. If we have evolved from ancestors with well-balanced and appropriate social behavior, why do most of us frequently deny some aspects of our nature to support an image of the way we want others to see us? Why do we believe we have to impress the world by wearing sheep's clothing to disguise our wolves?

Wolves in Sheep's Clothing

Denial of our dual nature holds the potential for producing a Dr. Jekyll and Mr. Hyde personality, an unhealthy split and lack of integration that deprives one of a full life. "Dr. Jekyll and Mr. Hyde" is the epitome of a wolf in sheep's clothing. This is the kind of man referred to by Shakespeare in *Measure for Measure*:

> *But man, proud man, drest in a little brief authority,*
> *Most ignorant of what he's most assured,*
> *His glassy essence, like an angry ape,*
> *Plays such fantastic tricks before high heaven*
> *As make the angels weep.* [7]

Mr. Nelson was a well-respected language teacher in my (Dolf's) high school. He was always well dressed and groomed, demonstrated impeccable social skills and was considered a pillar of our small-town community. He and his wife lived in a fine house adjacent to the river and had an active social life. Important people seemed to like him. One rainy November day, he left school in the middle of the day, went home, wrote a letter to his wife, walked to the river, undressed, neatly folding and stacking his clothes and placing his watch on top of the pile—then, entered the river and drowned himself. The event was the talk of the town for days. To most people, his drastic act of self-destruction was inexplicable and very tragic. The sentiment was that one of the finest men and teachers in our town was sadly and prematurely dead.

As a twelve-year-old student, I believed I knew why Mr. Nelson had taken his life. I knew he was a man who pretended to be who he was not. I can still see the hateful, cold eyes with which he looked at me. He seemed to delight in making students suffer with embarrassment when any of us made a mistake. I suspected that he knew I could see him as he really was—a sneaky bully who hated his students. It was as if the two isolated solitudes in which he and I lived collided. I disliked and feared him.

I somehow felt freed by his death, but also felt isolated, because the chorus of eulogies was overwhelming and the stamp of approval on his life so firm that my deepest feelings about him were invalidated. I felt certain that he had derived pleasure from making his students squirm and that he was intentionally cruel to us. I could not understand why others seemed to only see his outer shell. The worst part of this experience was my fear that maybe I was wrong—that something was wrong with me. I saw him as an evil man, but if no one else agreed, could I be the one who was evil? This bothered me for years. I finally told my mother of this dark shadow in my life. She admitted that she also had intensely disliked Mr. Nelson. This helped me to trust my instincts.

Our personalities can also become distorted by trying to live in the extreme aggressive end of the continuum to cover vulnerability. Ben, a relative of Paddy's, spent his life appearing tough and critical. He was difficult to love because we were never sure when he would lash out verbally, and sometimes physically, at any of us who were frequently around him. We knew that he had had a difficult childhood, living with an overly critical father, who constantly berated him. Yet, that didn't make it easier to accept his unjustified criticisms. Out of his fear that others would think less of him if he ever appeared weak or vulnerable, he could never ask for, nor accept, help. Then, in the last few months of his life, he became critically ill and almost helpless. Three weeks before he died, a miracle happened. The monster side of his nature seemed to melt away. It was as if he accepted his true vulnerability when death was imminent; his need to be loved and to express his love became his driving, dominant need. I spent many hours with him the last week of his life and witnessed a transformation, which others, who were not in his hospital room, find difficult to believe.

The night before he died, a group of seven deer came to frolic in the snow outside of his window. I experienced a strange sensation that the deer were coming to accept his soul, when, almost simultaneously, he murmured, "My friends have come for me."

Rather than being a wolf in sheep's clothing, perhaps he had been a deer in wolf's clothing.

As We Are

Why should we try to impress others by trying to act superior or different? Most of us do not want to be impressed; we want and need to be loved *as we are*. Simultaneously, we remain fearful that this is impossible for we are conscious of our individual potential to harm and destroy. We recognize that the world is dangerous because it is composed of other human beings—just like us. Since the beginning of those first life forms, nature has operated on two seemingly opposing principles: construction and destruction. Nature creates through competition and cooperation of molecules. We, as advanced organisms, also have the potential for constructive and destructive behavior. When we are hurt, angry or threatened, we have felt the urge to harm those who have triggered our painful feelings. Our saving grace is that we also have the desire and a natural need to help others, to care, to heal and to love. We have the conscious intelligence to choose how we will behave, based on our wondrous capacity to deduce, or imagine, if you will, the most likely consequences of our behavior.

When Mother Theresa was asked why she spent the later years of her life rescuing the infirmed and indigent in India, she replied that she knew what she must do the moment she recognized a "little Hitler" inside herself. We each need to recognize our internal apes, angels, devils and saints. We decide several times a day which of these aspects of ourselves to nurture. Our conscience judges our decisions as right and wrong, good or bad. We choose to feed the appropriate wolf for each situation.

Conscience

When I (Paddy) was a child and felt confused by having to make a simple choice, such as which friend to invite over or what I really wanted for my birthday, my mother would always say, "Choose what is right" rather than tell me what to do. It used to frustrate me and I would argue that I didn't know what was right. She would respond, "Your conscience will tell you." I never wanted to admit that I wasn't sure what a conscience was, but whatever, I would go out to my swing, tied from a limb in a large oak tree in the backyard, and wait for my conscience to talk to me. After a while of swinging and allowing my options to float around through my mind, I would realize that I knew what I really wanted to do. I then would come in to tell her. I could tell by her expression whether or not I'd made the "right" decision. When she did not look happy, I would change my decision because above all else, I wanted to please her to be assured of her love. Our parents are the original guides through the complex set of our innate feelings of right and wrong from which we extract our conscience, but as we mature, our conscience develops its own integrated program, based upon our memories (including the consequences of past decisions), our degrees of social relatedness, self-reflection, maturity, and our emotional stability. Emotions can be fickle, but they are our most honest internal indicator of our personal truth in any given moment— even when their expression may be socially inappropriate.

Although we cannot put a scientific finger on a physical part of our brain that is called a conscience, we are all consciously aware via the experiences of our lives that deep in our consciousness sits a moral judge. Possibly this judge is best identified by Gandhi when he referred to "the still small voice within"—to which he listened prior to making a decision or taking any action of consequence. For the purposes of this book, we will define conscience as that part of our consciousness that helps us determine which wolf to feed, or which aspect of our nature to trust in any specific situation.

Conflicts remain an integral part of us. We want the freedom to do what we want and seek the safety and protection of community simultaneously. This can often be visualized as a conflict between wants and "shoulds." The "shoulds" are imposed by our families, communities, religions, and other organizations that set standards and rules to help us manage our lives so that we will not disrupt the group. It is easy to allow our conscience to be controlled by groups to which we belong, however, there may be times and situations in our lives when our private moral judge—"our individual still, small voice"—whispers that the collective is on the wrong track. A conscientious objector knows this dynamic well. When we experience a "dark night of the soul," i.e., allow ourselves to be dipped into the hell of a moral dilemma that stretches us to our personal limits and often pits us against the group, we are forced to rely on our individual conscience.

The wonderful allegorical tale, *The Emperor's New Clothes*, is the classic example of how the conscience of a group can be ill formed by wanting to please a power figure, which is a major problem re the politics of war. When the young child in the allegory blurts out that the pompous emperor is naked, he cracks open the group conscience and drives home the truth that had been previously denied by the adults.[8] The child represents unadulterated truth. Truth does seem to emerge eventually. When it does, our conscience smiles, for it has usually haunted us with a deeper truth than we wanted to acknowledge in our efforts to gain approval and love. When love is not based on truth, we destroy the most precious gift life allows.

We are usually taught that jealousy, hate and feelings of superiority are bad, but at some point in our lives, most of us will harbor feelings of racial, sexual or religious superiority that spring from a deep-seated xenophobic level of human nature. Because the expression of these feelings is so strongly condemned in our society, we do not admit them, but we know they are present. This often causes us to doubt whether or not we deserve to be cared for or loved. The minute we recognize that we need love, we are hounded by the fear of losing that love. Denying what we fear within ourselves can lead to sudden outbursts of racism, sexism, homophobia or other violently antisocial behaviors. This dynamic is complicated and requires careful consideration by each of us. We need to be able to recognize our dichotomies, understand our conflicts, and know that all emotions, whether we perceive them to be positive and negative, are acceptable. All behaviors are not. We are not able to choose our emotions, but we are able to consciously choose our behaviors. Our personal conscience is our guide and our taskmaster.

To understand ourselves, we must stop hiding our true motives for behaving as we do, and accept that we all have antisocial aspects to our nature. It is easier and more comfortable to project these darker aspects of our nature onto others than to acknowledge them within ourselves. Unless we remain vigilantly aware of our internal conflicts, we cannot be true to ourselves. Bo Lozoff, founder of the Human Kindness Foundation in North Carolina, frequently quotes wise advice from one of his spiritual mentors: *If your goal is to understand the universe, you will understand nothing. If your goal is to understand yourself, you will understand the universe.*[9]

References:

Q. Shakespeare, William. from *Hamlet*. In Bartlett, John, *Familiar Quotations, 13th Edition*. New York: Little, Brown, and Company, 1955.

1. Sagan, Carl. *Billions and Billions*. New York: Ballantine Books, 1998.

2. Popper, Karl, *The Self and Its Brain*. New York: Springer Publishing, 1985.

3. Boehm, Christopher. *Hierarchy in the Forest*. Cambridge, Massachusetts: Harvard University Press, 1999.

4. Siegel, Daniel. *The Developing Mind*. New York: Guilford Publications, 1999.

5. de Waal, Frans. *Good-Natured: The Origins of Right and Wrong in Humans and Other Animals*. Cambridge, Massachusetts: Harvard University Press, 1996.

6. Freud, Sigmund. *The Basic Writings of Sigmund Freud*. Translated and Edited by A. A. Brill. New York: Modern Library, 1938.

7. Shakespeare, William. *Measure for Measure*. In *The Works of William Shakespeare, Complete*. Ed. by Walter J. Black. New York: Black's Readers Service, 1944.

8. Anderson, Hans Christian. *The Emperor's New Clothes*, adapted by Stephen Corrine in *Stories for Seven-Year-Olds*. London: Puffin Books, 1964.

9. Lozoff, Bo. *We're All Doing Time*. Durham, North Carolina: Human Kindness 1985.

The Power of Passion

The natural man has only two primal passions, to get and to beget.
~ Sir William Osler

George remembers himself as a teenage gargoyle and as a slave to his passionate adolescent fantasies, which made him feel he'd become an alien to his former self. At age seven, he began his boarding school education, where he excelled athletically and academically. He knew that he was "good at things," so maintained fairly high self-esteem for the next several years. During this time, he was aware that girls were around, but claims that he had little use for them, as they were seldom much good at sports. Before he left home, his mother had explained the facts of life, thinking he should know where babies come from, but he thought she certainly had things screwed up. She had already proven a very dubious authority on important matters, such as sailboats, cars, and airplanes. His reflection on this conversation was that he couldn't imagine enough males would think of doing such a thing with a girl to propagate the earth.

Fast forward to age thirteen when puberty changed every aspect of his earlier life. He claims that he resembled Ichobod Crane, at 6'2" and weighing only 140 pounds. He was so gangly that he tended to clank when he sat down in wooden chairs. His physical image was further dis-enhanced by a world-class case of acne. He felt that his gargoyle appearance, awkwardness and lack of social graces actually made girls run for miles. His self-esteem and his academic grades plummeted. The only thing on the rise was a massive, tantalizing preoccupation with sex, but he knew absolutely nothing about females or love, except what was derived from copious reading of Victorian fiction and Arthurian romance. He was falling passionately in love with his fantasies, fueled by adolescent hormones.

At age eighteen, having never had a kiss or a date, he was drafted. It was the summer of 1945, during which casualties of WWII were running 20,000 per week, and the war was predicted to last nine more years. George was sure that he would never live to have that first kiss, not to mention the other passionate experiences about which he'd been reading and dreaming. During his basic training, which was not difficult for him as he was a well-trained athlete, women remained the impossible dream. After completing seventeen weeks of basic training, he was given his first week's leave. He donned his Army dress uniform and headed for the bus station. To his amazement, a pretty girl in the adjoining bus seat began taking an active interest in him; it seemed that his uniform made him glamorous in her eyes. Although he remained un-kissed as she lived hundreds of miles further on the route from his bus departure, his ego had received quite a boost. Following the triumphant end of World War II, females still found military uniforms glamorous symbols of male protection. Not long after, George was stationed in Paris, where American soldiers were very much in fashion. A beautiful, intelligent haute bourgeoisie French girl found him interesting indeed—and rectified matters as his fantasies materialized into reality.

Was it only raging hormones, or did the war and the uniforms intensify the situation—especially because Americans in uniform were "the heroes" that year? Erotic love and war seem to be the situations that impassion us more than any others, as Tolstoy's Count Rostoff reminds his son in their discussion on passion.[1] Certainly, erotic love adds a dimension of meaning and of energy to our individual lives that seems to be lacking when we are not in love. Competitions, even those as deadly as war, raise the energy level of groups and of nations. Each of these situations shares elements of risk—both emotional and physical—and an opportunity for victory or defeat. Are these elements necessary for us to experience passion? Let's investigate more thoroughly the anatomy of passion.

Passion

Passion is the internal experience of emotional intensity—in short, it is any emotion on fire. Emotions verify our existence and increase our sense of aliveness. Our most fundamental psychological need is to experience our aliveness. Therein lies the power of passion. The late Howard Thurman, a renowned American theologian, urged us to follow our passions if we want to experience the ecstasy of being fully alive:

Don't worry too much about what the world needs. Ask yourself what makes you come alive and do that, because what the world most needs are people who have come passionately alive.[2]

The ironic fact is that in living the full experience of ecstasy, we are equally vulnerable to the full experience of agony. Once the door is opened to emotional capacity, we are not able to choose what we feel. As the tin man in the Wizard of Oz cries, "I know I have a heart because I can feel it breaking."

George's raging hormones, a genetically programmed aspect of adolescence, came naturally. What to do about managing them did not. Most young adolescents are grateful that masturbation is not a legal crime, yet what we do alone to manage passionate feelings rarely proves as satisfactory as what we can do in concert with another individual. Passion seems to push us to go beyond ourselves, which almost always involves risks, sometimes deadly serious ones. Thoughtful and rational people find themselves torn between the rush of ecstasy that comes with a passionately emotional experience and their rational inner voice warning them to cool down and consider what the consequences of reacting only on emotion might be. We run the risk of rejection the minute we reach out to another, so expressing passion often requires an enormous amount of courage.

Then again, passionate outbursts can happen with complete spontaneity. When Roberto Benigni won an Academy Award in 2000, for his performance in *La Vita e Belle* (*Life is Beautiful*), he was so overcome with passionate appreciation that he jumped over the rows of seats, with people seated in them, between himself and the stage, screaming thank you's in several languages. Because these people knew him and could relate to his emotions, they laughed, cried and applauded, but had a complete stranger rushed into the auditorium and exhibited the same behavior, he may well have been arrested as a madman. Therefore, context is an important aspect of what makes passion acceptable or unacceptable to others. Benigni's passionate acceptance of his award will be remembered long after we forget who won other awards that evening, because passion is unforgettable and contagious.

The contagious aspect is obvious at most professional and collegiate ball games with competing teams decked out in uniforms of their team's colors and with scantily clad cheerleaders cavorting about to add a touch of sexual stimulation to the already energized crowd. Sometimes these crowds are not only screaming and shouting, but can become violent. If an alien from another planet happened to drop in on this scene, he may correctly think he is observing a frenzy of human madness. Is a touch of madness inherent in passion?

Academy award ceremonies and sports events are rather innocuous situations that may appear to contain a touch of madness; however, killing and war represent passionate madness in its extreme forms. They result in suffering and endless pain that can rarely be reconciled. Why are supposedly rational and sapient beings willing to accept such risks? Why are we willing to throw ourselves passionately into situations, which in an evolutionary sense, are liable to reduce our biological fitness? Would we not have evolved greater control over our passions, unless our need to experience them offered rewards?

The original translation of passion from Aramaic is *suffering* and came to us from the Greek word, pathos, which is used to describe the darkest hours of Jesus' suffering at his crucifixion. The emotion triggered by this experience was so powerful that Christianity, one of the dominant religions of the world, was based upon the event. Over time we began to correlate passion with intense love, for it was because Christ loved intensely that he was willing to suffer intensely. Passion pushes us to our emotional edge. Often passion is blamed for causing us to lose our ability to be rational—for pushing us over the edge. There is sometimes a gap between what we feel we want to do with passionate feelings and what others want us to do—or will allow us to do without punishing us. In this gap lies the dilemma caused by the power of passion. We believe it is time to take a serious look at how this conflict is played out in our brains.

Our Emotional Brain

Emotions are often referred to as our heart's way of knowing. The truth is that we actually have an emotional mind, which preceded our rational mind in evolutionary development. The brainstem is the most primitive part of our brain. Our brainstem is not able to think, or even to learn how to think, but it is programmed with a set of regulators that keep us alive and automatically reacting in ways that ensures our survival. This part of our brain is sometimes called our reptilian brain. It allows a snake to hiss when threatened by an attacker, just as it causes us to scream or jump when we feel threatened or surprised by an intruder, even though it may be someone we know well.

In his book, *Emotional Intelligence*, Daniel Goleman, psychologist, writes that we have feelings first and thoughts second, as the emotional mind is quicker than the rational mind. It springs into action without pausing to contemplate a situation. Goleman's research suggests that a tendency to act is implicit in every emotion.[3] This is more often obvious in the behavior of animals and young children than in grown-ups. It's easy to tell when a puppy is happy, as it will wag its tail and tends to wiggle all over; likewise, if the animal is fearful or sad, it will tuck its tail between its legs and sulk. When a young child feels angry, hurt, or frustrated, he is quick to yell, cry, kick or bite. When a young child feels happily surprised and thrilled over a new toy, she will squeal with delight and often hug the person or object that has elicited the emotion.

Adults, with our more developed rational minds, tend to be more circumspect in expressing our emotions freely *until* we have assessed the level of risk involved. In most cases, we will hold in, or repress, intensely

43

passionate feelings until we find a safe place to express them without being accused of having poor impulse control. However, in dangerous situations, we sometimes owe our lives to the instant reaction of our emotional mind, which moves faster than our analytical mind and allows us to duck for cover when we hear an explosion or notice a rapidly moving object coming towards us—anything from a golf ball to a Mack truck.

On the other hand, our rational brains can also be life saving. I, Paddy, recently took a dive out of an airplane at eleven thousand feet above the ground. Of course, I was strapped to a parachute, which was not to be opened until I reached five thousand feet, indicated by the altimeter strapped on my wrist. This was to allow me the thrill of free falling through space, of living for a very few minutes at the edge—not knowing whether the chute would open. Suddenly, I was aware of an intense volume of air that was rushing down my throat with such force that I couldn't breathe. I'd forgotten that the instructor's last words before I tumbled out of the plane were, "Keep your mouth closed!" At the moment I was sure I would die, my rational brain kicked in—I closed my mouth. Able to breathe again, I opened the parachute on schedule and floated blissfully to the ground.

The capacity of my rational brain to "kick in" under duress illustrates one of the selective forces that led to the evolution of our rational brains. This ability is frequently observed in nature. When a polar bear charges a hunter who comes in close to the bear's kill, the hunter can usually dissuade the bear from completing his charge by firing his gun in the air. The bear will have the emotions of both fear and rage, but the loud noise of the shot will raise the level of fear to overrule the rage. Fear is one of our strongest emotional states, and therefore has powerful control over our actions. It is also possible that the bear, having developed a certain level of rational thought, has learned about the efficacy of humans with guns, and will run rather than continue the charge. The more cognitively advanced gorilla in the same situation will begin a charge, but in a conscious, intentional bluff to frighten the hunter, but stop the charge before contact can lead to physical harm. As for the hunter with the gun, we can only hope that he, too, will behave rationally—fire the gun in the air, walk slowly backwards away from the animal, and thereby spare all lives. Unfortunately, this is unlikely behavior for some hunters.

The Nose Knows

It might come as a surprise that our emotional life began in the olfactory lobe of our brains, with the sense of smell, like the vole in chapter one that smelled the fox, felt fear, and changed his route in his search for food. Every animal, including the human variety, has a signature smell, caused by volatile substances, some of which have evolved to become pheromones. Pheromones are the chemical substances produced by an animal, which elicit an adaptive response in other individuals of the same species. Our ancient emotional brain's first job was to analyze smell and sort it into categories: edible or toxic, sexually available or enemy. The next level of cells then sent a message for action to the reflexes: bite, spit, approach for copulation or battle, play dead, or run for your life.

After early life survived and reproduced through this olfactory stage of brain development, another level of cells developed into what we know as the limbic system, which formed a ring around the brainstem and increased our emotional repertoire. The hippocampus, a storage site for memory, and the amygdala, an almond-shaped cluster of interconnected neural circuits located above the brainstem, are key components of the limbic system, which in turn gave rise to the rational cortex and neo-cortex. As the limbic system evolved in mammals, it took on the tasks of making distinctions among smells, comparing them to past experiences and making decisions about how to behave the second time around. In other words, learning began to take place, which allowed mammals to change a course of action rather than only re-acting automatically. In evolutionary lingo, the "nose-brain" had evolved into a thinking brain.

I See You

Many millions of years ago, when our primate ancestors became diurnal and arboreal, they also became more visible to each other and began to develop an increasing dependence on their sense of sight, combined with their already developed sense of smell. At some distance, these arboreal ancestors could distinguish mating signals in others of their species, which added visual stimulation to their sexual repertoire. During the gazing at these displays, the brain is flooded with phenyle-thylamine (PEA) and natural amphetamines, which produce feelings of euphoria and passion. The pituitary gland then secretes oxytocin, which anthropologists refer to as "the cuddle chemical," as it seems to produce a need to embrace the other. Via these biochemical developments, gazing and hugging became a part of our ancestors' love behaviors—and were thankfully passed on to us.[4]

Modern humans seem to be much more aware of the visually recognizable attributes of potential partners than of olfactory ones; however, we still carry a latent ability to recognize and assess people around us by scent. Several years ago, I (Dolf) was asked to be Father Christmas for the nursery school where my four-year-old son spent three mornings a week. Clad in a Santa Claus outfit stuffed with pillows, a beard obscuring my face and carrying a sack of presents, I entered the classroom, shouting, "Ho, Ho, Ho." The din of twenty excited children instantly died down to near silence, except for one small voice, which piped up, "There's my daddy." As Santa entered the room, his little boy, who was approaching him, stopped dead in his tracks and started to step back, no longer completely trusting his visual and auditory senses. I also stopped, both amused and a bit worried that my son might be frightened. Then, he walked forward, stuck his face in Santa's crotch, sniffed, and with a big smile exclaimed, "It is my daddy!"

I also have a highly sensitive olfactory sense. After my wife's two pregnancies, I had no trouble identifying pregnant women by their unique subtle scent. This led to an embarrassing situation once when I asked a friend when her baby was due. She became agitated and said that she had not yet told anyone and was upset that she might already be showing signs of gaining weight. Having to explain that she smelled pregnant was not a good way to continue a polite conversation. Since then, I've kept my olfactory skill to myself. Most of us, however, do respond subconsciously to pheromones exuded by others, which can illicit feelings of sexual attraction and passion. Yet, the abundance of sexual material used in advertising and the accessible visual pornography in most cultures today (2010) indicate how we have moved towards more visual stimulation.

The Seat of Passion

Although we feel as if there must be a god of passion that catches us unaware and tosses us about like leaves in a whirlwind, it is actually that small almond-shaped cluster of interconnected neural pathways that stimulate us to passionate heights—and depths. The workings of the amygdala and its influence upon the neo-cortex intensifies our emotions and causes us to fall passionately in love (or lust), to turn anger into rage, sadness into severe depression, or enjoyment into ecstasy. Our "seat of passion" is our amygdala, not our heart. But, romantically speaking, "I love you with all my amygdala" just doesn't make the cut.

What this "seat of passion" actually does is to coordinate all incoming information from our sensory organs, and then very quickly sends it to our hippocampus, which codes the information almost instantly. Before we can "think about it," we're only feeling and reacting. Our emotions keep us honest, although this is not always what we would rationally prefer. This seems to be especially true when we have a negative reaction to a situation but do not want others to know how we feel or when we do not feel the way we rationally believe we "should" feel. For instance, when we say, "I am not angry" or "I don't care, it doesn't bother me at all," our tone of voice, facial expressions and bodily responses are more accurate indicators of what is going on than our words. It is easy to lie with words, but emotional reactions keep us internally honest. The power of an intense emotional sensation can cause us to blush or to blanch, feel breathless or nauseated, which are usually giveaways that an emotional truth is in conflict with what we are trying to convey. A passionate feeling can so totally overtake us that it seems as though the rational part of our brain has taken a coffee break.

When the amygdala has been severed or removed in animals, as well as in humans, all emotional reaction disappears. In the recent past, neurosurgeons removed the amygdalas in a few epileptic children to prevent severe seizures. The dramatic result was that these children became void of feelings—unable to cry their tears or laugh their laughter. They lost their fears, their honesty, and their ability to love. Their lives were irreversibly diminished. The procedure was quickly abandoned.[5]

The Passionate Stages of Life

Sam Keen, psychologist-philosopher, in his book, *The Passionate Life*, presents a useful model to explain our need for passion and how it develops in our lives. Keen constructs a life-map that traces the development of passion throughout the course of an *ideal* lifetime. He arbitrarily divides our passionate development into five stages: the child, the rebel, the adult, the outlaw, and the lover.[6]

Stage 1: The Child's Being

According to Keen, the primary motivation of children is to stay passionately attached to their caregivers—to secure love from the primary caregiver, usually the female, and to be accepted and loved by the male caregiver. When this is done in a healthy manner, the child develops suitable dependency, basic trust, openness, wonder, obedience, curiosity, and the enjoyment of intimacy from being appropriately loved by childhood caregivers. This allows the child to enjoy passion without fear of rejection or of being punished. This is an ideal scenario, for, in real life, most of us were punished at times for expressing what others considered inappropriate expression of intense emotion.

Every "nice" emotional state seems to have a "bad" counterpart, such as happy and sad, contentment and anger, etc. Although all emotions are acceptable under certain conditions, when the emotions considered more negative by our culture are intensely felt by children, they are usually told they *should* not feel that way. This seems to hold especially true for little boys who are taught that they should not feel afraid while little girls are taught they should not feel angry. Needless to say, we each feel what we feel. Being told that some feelings are not allowed is very confusing—and also presents an impossible demand. This conflict between honest emotional reality and acceptable emotional expression begins as early as age two or three years, depending upon the emotional boundaries of our caregivers. Trying to resolve this conflict between what we call human nature and the expectations of society stays with most of us for the remainder of our lives.

If we have siblings, childhood also introduces us to passionate jealousy—over parental love and attention, or even over toys. We learn to argue, to fight, and manipulate to obtain what we want. Sometimes we are punished, but sometimes we are rewarded by achieving what we want, so these emotions are reinforced, even though we are usually taught they are "bad." Perhaps it is the safest way to ease into learning that we can love and hate the same person and to learn how to handle our more negative emotions. They are a part of our nature, but how to manage them requires experience.

Think of the moments in your early life that elicited passion. Were you allowed to freely express your feelings? These moments are usually unexpected, totally consume us with emotion and are never forgotten. I (Paddy) remember the first time I saw the ocean; the memory is as clear as if it were only five minutes ago, rather than over 60 years. I was six years old and our family had recently moved to a coastal North Carolina town. My father took my little brother and me to "meet the ocean." He told us to close our eyes while he held our hands and led us over a row of high sand dunes. This was a bit scary, but since I totally trusted my father, I was equally excited about what he might have in store for us. When we reached the top of the dune and he said, "Okay, open your eyes and meet the Atlantic Ocean," I was so totally awed by the vastness, beauty and power of the sea that I laughed and cried at the same time— which is not unusual when we are emotionally overwhelmed. What I didn't rationally understand at that time was that I fell in love. That love has dwelled deep within me for over sixty years. I've experienced hundreds of wonderful, fun-filled, wild and crazy times in many oceans, while I've also encountered its destructive powers—to wound and to kill. I've learned to respect and to fear it, but when I am away from it for over a year, I long to be close to it. And, even now, when I first glimpse the ocean after a long separation, I still feel that same sense of wonder, as if I were seeing it again for the first time.

My father was a man of very deep emotional capacity and I am both blessed (and sometimes feel cursed) that I inherited that tendency, for we do inherit a tendency to be either more passionate or more placid from our parents. Our inherited tendency is then encouraged or discouraged by all of our major caregivers' and by the limits they set for us.

Stage 2: The Rebel Temperament

The second passionate stage is that of the rebel, who defies authority and establishes identification with a peer group, which replaces identification with our parents. The rebel temperament is independent and self-conscious, seeking new ideals and choosing new heroes and heroines. In this stage, which normally occurs in adolescence, we tend to idealize and adore our boyfriends and girlfriends, driven by heady fantasies of what romantic love offers, such as our George in the beginning of this chapter. The healthy rebel learns to doubt, criticize conventional wisdom, and test the limit of boundaries and rules. Rebels tend to feel every emotion as passionately as possible and could care less about being rational or behaving that way.

Adolescent interactions may be extremely romantic and loving, but they may also be competitive and violent. We enter the rebel stage of our lives with an individual status provided by our caregivers and their social set; however, as we extend that world, we are always challenged by other adolescents and by our rapidly changing bodies, emotions, and freedoms. We have to reestablish our status by proving our abilities to compete, to lead, to control and to take risks. These risks run the gamut from becoming passionately involved in relationships, social causes, and athletic competitions to defying authority, taking drugs, engaging in irresponsible sexual experiences and worse. During this stage, we seem to be fearless and almost unaware that our negative behaviors can have negative consequences. Most adolescents, regardless of their behavior, live under a "personal fable" that they will not end up in jail, become pregnant, or be killed.

The sad reality is that this is a stage at which destructive things do tend to happen frequently. Research from scientists at the National Institute of Health indicates that the "executive" part of our brains—the part that weighs risks, makes judgments and controls impulsive behavior—isn't fully mature until after the mid-twenties, and is therefore not adaptive to forming a safe society. One of the most negative aspects of this phenomenon is that adolescents are the ones most often called upon during wars to defend their country. They are encouraged to be aggressive and to inflict lethal violence towards "the enemy." Even though it is well known many do not survive war, the ones who join the military forces are made to feel enormously vital and important. When the war is over, those who survive return home as "heroes" and "heroines" and are praised for their aggressive abilities. This develops a conflict-ridden mentality, which can lead to a repression of gentleness to gain the rewards given for the display of more aggressive behaviors. This is not only a phenomenon of war, but of athletic competitions and other situations where victory goes to the aggressor. We have not yet figured out that balancing our energies, choosing our passions with some degree of rationality and concern for consequences of our behaviors are necessary for maintaining a stable civilization.

Those of us who have endured past adolescence often look back upon this stage as one of passionate turmoil, a time of testing our limits, a time to be remembered—but, hopefully, not to be repeated.

Stage 3: The Adult Personality

After completing the work of the raging hormone adolescent rebel, we ideally move into a more rational adulthood, wherein we seek the stability of membership in our community and often repress our most passionate feelings—at least publicly. During this stage, many of us take marriage vows or make another strong personal commitment, create a home and family, and invest energy into raising our young or making other contributions to society. The healthy adult personality demonstrates responsibility, can delay gratification in order to give to others, becomes more predictable and consistent, obeys laws, and respects authority when it is just and reasonable. By this time, we have learned that living our every passion can have damaging consequences. However, passions can be acceptably channeled into sports, hobbies, supporting charities, doing volunteer work for religious and political organizations and into our careers. We are prone to praise passion when it is demonstrated in ways that help maintain or improve our communities or culture, but condemn it when it seems to disrupt the structure of our organized society's value system.

Tim, Paddy's husband, was a passionate builder of model planes as a child, then became a pilot as a young adult and also discovered the joy of soaring gliders. In his retirement, he keeps his passions alive by building and piloting power planes and gliders. After over thirty years of marriage, as the passion of sexual experiences began to ebb, I realized that I must learn to fly if I wanted to maintain any passionate connection to my husband—at least, it would keep conversations lively. Although there were moments when I truly thought the endeavor would destroy our marriage, or kill me, it definitely has paid off. Some of our most passionate connections occur when we are piloting a plane together. "Your airplane," he says, as he gives me control of the plane—and I feel a sudden sense of power, pride and an appreciation of his passion, which I now share.

It's important to realize, however, that the adult does not suddenly become a law-abiding, selfless contributor to a stable community, or a paragon of virtue. In fact, most men and women walk a tightrope of conflicting desires. These desires vacillate on various continuums—from selfish to unselfish acts, from rational to irrational decisions, from high risk to secure investments and from engaging in intense passionate experiences to accepting the status quo, even when bored. Adults are never exempt from conflict for extended periods of time, yet adults are more aware of the possible consequences of their decisions and passionate involvements—pertaining to relationships, careers, finances, hobbies and other interests. By adulthood, we have usually experienced "both sides" of several passionate experiences—the agony and the ecstasy. Sometimes, when considering taking a risk, adults will think "no pain, no gain," yet, there is usually a limit to the pain we are willing to endure for a prize that cannot be assured. Most adults hold these limits tightly in place until the outlaw self begins to stir forgotten passions and dreams.

Stage 4: The Outlaw Self

The outlaw self often manifests following a stable period of relatively low passionate involvement. It is as if our psyches become bored and tempt us to seek more autonomy and independence. When bored, we tend to take more risks in order to experience a more passionate life—not necessarily at the expense of severing established intimate connections. Although, separations and divorces are more apt to happen at this stage than in the previous stage. The tasks of this stage, which usually occurs during middle age, are to re-own the darker side of our natures, without seeing ourselves as evil. This gives us the courage to drop our illusions and defense mechanisms, to become more honest and able to gain more insight into what is required for us to stay passionately alive. It correlates with Carl Jung's idea that we cannot develop the capacity to understand who we are as unique individuals or to reflect on the deeper meanings of life until we are past age forty. A basic requirement for wisdom, according to Jung, is to have experienced a few dark nights of the soul.

Anna, a moderately happily married woman in her mid-forties, had reached that time in her life when her three children were out of the nest; she was restless and moving into her outlaw stage. She went back to college to complete a degree she had begun over twenty years ago. Being back on a university campus was stimulating, challenging and fun. It was even more fun when one of her professors seemed to take a special interest in her. He was also married. This made her feel safe. Before long, her dreams and fantasies began to include this professor, but she would engage her rational mind and tell herself that her fantasies were utterly silly. Her internal wolves, one representing her titillating fantasies and rich emotional life and the other representing her more rational life were engaged in constant, heated debate. When the rational wolf scored points, she would enter the classroom determined not to make eye contact with the professor. The harder she tried not to look into his eyes, the more her eyes seemed to take sneak glances to see if he were looking at her. When he was, she felt thrilled, but when he was not, she felt miserable. Knowing she was being foolish, she would catch herself wondering if he had lost interest or if he were connecting with another female student.

He stopped her one evening as she was leaving the classroom and asked her to go to the campus center for a cup of coffee. She intended to refuse, but surprised herself by hearing her own voice say that she would love to go. They ended up in his office having sex. She called me (Paddy) later that night, making an attempt to remain calm, while telling me that she was losing her mind. It took only a few more sentences before her tears and story poured forth. Her biggest concern was how she had ever allowed herself to get into such a mess when her life had been "perfectly happy" prior to her taking this particular class with this particular professor. In a therapy session the next afternoon, she appeared much calmer as she tried to convince me that she really was going to be fine for she had decided never to go back to his class, to accept a failing grade, if he dared give it to her, and that she didn't give a damn. I only looked at her. She began to sob while saying that she hated him and hated herself—and hated her whole damn life. "The kindest thing you could do for me," she sobbed, "is just give me hemlock."

We both laughed. I said, "I understand." She continued to cry, but without the self-condemnation, which she seemed to be feeling a few moments earlier. When we think we may be losing our grip on rationality, or "going crazy," if there is another person who admits to having been in that same mental, emotional space, that person can become a life-line back to sanity.

Were Anna and her professor caught in just another frenzy of human madness? Perhaps. Yet, this true story has happened thousands of times with many variations on the same theme—the powers of passion that can overwhelm us. It is most apt to happen during this outlaw stage when we are searching for a way to reconnect passionately with life.

Stage 5: The Lover's Spirit

The final passionate stage of life is that of the inter-dependent lover, in which we trust ourselves to love and be loved without the neurotic interfering with the erotic. At this stage we have developed empathy and are moving into compassion, which encompasses the ability to accept differences without trying to control or manipulate others. We have choices—to choose to be loyal through the bad times and to make allowances for weaknesses, since we have learned there is no perfection. However, this stage does not require foregoing passion. The process of sharing life physically, emotionally, sexually and spiritually over years can reward us with an unsurpassed depth of passion.

Few of us ever achieve consistent mastery of this fifth stage. If we did, we would live in a far more peaceful world, with less divorce, fewer crimes, and happier homes. This is not to imply that all relationships should last. There may be such an accumulation of negatives with so few positives that to continue a relationship would not be constructive to either partner, or to their children. When this is the case, it seems best to move into the more advanced stages of love with someone who is able to co-create more positive energy. Whether or not we feel energized (impassioned) by a relationship is a key component in assessing its health. When a relationship saps our positive energy and disallows natural spontaneity, it is time to assess the value of the relationship. If we conclude that we are not able to live honestly or to be whom we really are, or if our deepest values and passions cause constant conflict, it may well be time to consider other options.

It seems likely that with each new relationship, we move briefly back through each of these five stages. The good news is that once we have totally achieved the goals of any given stage, we do not have to spend significant time at the earlier stages. However, as Katherine Anne Porter has written, *Love must be learned, and learned again and again; there is no end to it.*[7]

Love and Limerence

In her book, *Love and Limerence*, Dorothy Tennov, a Professor of Psychology at Bridgeport University, explores how we get caught in inexplicable erotic passions throughout our lives. She defines limerence (a word not found in *Webster's*) as a mental activity, which is set in motion when we meet people to whom we are strongly attracted—to the point that we become obsessed by thoughts and fantasies of them and by hopes that they are also thinking about us, which is what happened to Anna and her professor, as well as to other millions of us.

Essentially limerence is another term for what most of us call "falling passionately in love." It often has little to do with real and lasting love, but offers a springboard of hope that a new love will be the answer to all emotional problems. We rationally know that no relationship is without inherent problems, but we cling to the impossible dream that this new one will offer unmitigated happiness. Stendhal writes that romantic love is like a fever that comes and goes independently of will.[8] The experience of falling into romantic love illustrates the powerful connection between joy and grief, ecstasy and agony, fear and hate—which can lead to crimes of passion. Several people facing criminal charges of abuse have pleaded that they could not be held responsible for their actions because passionate anger or jealousy had rendered them temporarily insane.

Evolutionary psychologists Martin Daly and Margo Wilson report in their book, *Homicide*, the poignant story of man who shot and killed his wife when he discovered she had been unfaithful to him. Immediately after realizing what he had done, he called the police. When they arrived, he was sitting beside her body sobbing and repeating, "I really loved you." These authors suggest that if in the heat of extreme passion, we could learn to wait for our blood to cool down before taking action, many of these "crimes of passion" could be avoided.[9]

When I (Paddy) first asked my father how I would know if I were in love, he smiled and asked if I thought I'd know if I'd been hit by a meteorite. I considered this and remember saying, "That could kill me." He laughed and said, "Well, the difference is that you only think love will kill you at times. At other times you realize it is the only thing that makes its pain worthwhile."

That bit of conversation took place about a half a century ago. Since then, I've blissfully walked into the paths of several meteorites. He was right—I'm not dead—yet! I've realized that each time I get whammed, I learn more about myself than through any other life experience. One of the most unique features of falling in love, or limerence, is that regardless of the number of times it happens, it always feels like the first time. "This time it *is* different," says the man in his sixties who has been through four divorces and untold numbers of love affairs. It really is different each time because a different person will elicit new responses from us. We learn a bit more about ourselves each time we commit to a relationship. One of the important things we must learn is that a relationship cannot be sustained on passion alone. Healthy relationships involve our hearts—and our brains.

Analyze This

The movie, *Analyze This*, offers us a pretty good analysis of what is going on when our need for emotional stimulation overrules our ability to behave rationally. The film portrays a man trying to live on the passionate precipice between security and danger. He needs to prove that he can break all the rules of society by leading a mafia ring, while simultaneously maintaining the image of an upright citizen. It boils down to his inability to resolve the inherent conflicts between his darker nature and his more civilized nature. He appears powerful, but is an emotional coward in that he cannot choose to live an integrated life.

This never works long term, as the conflict tends to take on a life of its own, causing us a level of stress that most of us are not able to sustain without breaking down—literally becoming physically or psychologically ill, as the character in the movie does. As deeply ingrained as our need to live passionately, there is also a need for homeostasis—the place where we can maintain comfort and security. When conflict is raging internally, it pushes us to a crossroads at which a choice must be made.

In one of psychiatrist, R. D. Laing's last lectures, he described his views on this phenomenon via a story of a client who had called him from a cardiac unit in a London hospital. The client told Laing that he had just had his second heart attack and needed to resolve a major conflict or die. He confessed that he had been leading a double life, had two wives, two sets of children, two residences—one in England and one in America. Laing said that the problem was a simple one to solve—make a choice. The man pitifully explained that he had tried to do that many times over fourteen years, but just couldn't stand hurting either of the women, as he loved them both and needed them both for different reasons. As only Laing could have done, he began to walk out of the man's room and called over his shoulder, "So die and hurt them all."

Within six months, the same man, who had just had his third heart attack, contacted Laing again. This time the patient was ready to make a choice. Laing required that his patient come clean and confess to both wives, at which point the patient said he'd rather die, for he was sure his first wife, the British and more proper one, would kill him anyway. Realizing that he was being somewhat controlled by his fear of her power, he chose his American, second wife, believing this choice would give him the greatest opportunity to enjoy what time he had left.

This story had its very painful moments for all involved, including Dr. Laing. However, with his expertise and patience, this case was resolved without bloodshed or deaths. Both wives admitted that they had been suspicious for years and felt relieved when the truth was out—even though it is unfair to say that everyone rode happily off into the sunset. The fears of the unknown are usually greater than the reality of what actually happens when we clean up our duplicity. Most of us function at a higher level when all the cards are on the table, than we do trying to untangle a web of mixed messages, or lies.[10]

Certainly, all such cases do not have happy endings. Disasters and tragedies happen even when we put forth our best efforts to solve conflicts and choose what seems the least painful for the most participants. However, when we are willing to deal honestly with our internal conflicts, we gain the best opportunity for a positive outcome. My son-in-law, Jimmy, recently cut to the chase when he asked: "Mom, why can't people learn to communicate honestly and to talk through conflicts?"

This sounds rational and simple, perhaps even naive. Why is it so difficult? Why do we push the limits of our passionate edge—to risk learning to fly, or dying? To begin to answer these questions about our own species, let's take a closer look at another species.

Jousting in Earnest

As a biologist, I (Dolf) have had many opportunities to observe animals in action; their lives, seen through human eyes, can be heroic, pathetic, funny and tragic. I have also had the opportunity to contemplate how their behavior has evolved and what that means when it comes to understanding human behavior in scientific terms. While working in the Canadian Arctic, on Banks Island, I witnessed a poignant example of the passionate life of muskoxen.

The day began as a typical Arctic end-of-summer day of mist, rain and sleet. It had rained all night, and by get-up time, it was still splattering down on the canvas above me. It took a great deal of fortitude to leave the comfort of my sleeping bag and face the world. Low, dark clouds obliterated the tops of the hills, and the wind drove the rain sharply into my face. The thermometer on the outer shell of the parcol tent told me it was a mean thirty-six degrees. Once the stove was lit, and breakfast was cooking, we started to plan the day, and decided to complete the final survey of the experimental grazing pastures. This meant hiking from camp, past the shore of Eider Lake, to and through the area of marshy polygons, and from there into the hills to the northwest. Donned in our raingear overtop of our parkas, we were more or less immune to whatever the elements had in store for us, but the weather worsened as rain became sleet and sleet became snow. By the time we reached the higher levels in the hill country, wet snow had covered enough of the pastures so that observations on the vegetation would not be worthwhile. The warm tent beckoned, and we decided to abort our day in the field.

Then, on our way down to the flat land below, following a small stream in a sheltered valley, we came upon a large herd of some forty-five muskoxen. We noticed a commotion in the herd. Most of the cows and their young were lying down in a loose cluster, seemingly oblivious to what was happening some fifty feet away. There, surrounded by a gang of very excited yearlings and younger bulls, were two magnificent, fully mature bulls. They were standing boss to boss, pushing hard, but despite their weight and effort, neither prevailed. They kept pushing, repositioning their feet, rubbing their heads together, but they were evenly matched. Suddenly, as if by some silent command, they stopped pushing and stepped back while staring at each other, and then, slowly, each bull started to walk backwards.

They continued to stare at one another as the distance between them grew until they were well over a hundred feet apart. I did not hear or see any signal, but as if released by the sound of a starting pistol, the two animals started running forward simultaneously, closing the distance between them and gaining speed by the second. They collided with a loud thud, which echoed among the hills. The impact was horrendous, their bellies flopped forward, their tongues hung out of their mouths, their backs seemed to buckle, their eyes bulged, and blood reddened their faces. Then they stood there again, boss to boss, with steaming hides, pushing, shoving, and breathing audibly. I wondered how their brains and spines could survive such a violent collision, but despite obvious physical damage, these two bulls hadn't yet had enough. The prize was too great to give up at this stage. They repeated this awesome jousting act five more times, before one of the bulls began to flag. He started his run a bit later and ran more slowly. On impact, he staggered and was pushed back. He was bleeding out of his mouth and breathed hoarsely. After one last bout, he turned around and walked off slowly into the empty landscape, leaving the herd behind. The victor, still with heaving sides and breathing heavily, stood there, surrounded by admiring yearlings. For the moment, he was in charge of the herd.

Initially, it is tempting to see this as a fierce and violent clash between males who seek access to the reproductive potential of the cows in the herd, but not a lethal combat. In reality, we have observed that many of the bulls that take part in this fight for herd dominance do not survive the following frigid winter. When we arrive each June at the Bank's Island research site, we always find a number of corpses of large, mature bulls who have died during the preceding winter. The corpses consist of skeletons ripped apart and chewed clean by wolves, foxes and ravens; sections of hide partly covered in hair; and a large pile of undigested stomach forage content. It is obvious to us that the animals had not died of starvation, and with a lot of calves and yearlings around, it is unlikely that wolves killed the bulls.

A wildlife vet established that most of these bulls had died of a disease closely related to plague, which seems to affect weakened individuals in the population. The potential cost, but also the potential prize, for entering the annual contest for a herd of cows is enormous. Failure will probably lead to death; success will make you the sire of up to thirty calves, but could still cause your death. Most bulls do not enter the annual contest seriously; they may challenge a herd bull briefly, but give up before sustaining any injuries. They usually live for years in small bachelor herds, finally entering the rutting fray when fully mature—even then with only a small chance of success.

In a world where cows and juveniles have to live in a herd structure for security, it is easy to see why natural selection favors a strongly polygamous breeding system. In the case of the muskox, natural selection has obviously favored males who persist to the bitter end in their drive for domination of a female herd. High hormone levels, a highly specialized physique for fighting other males, and the high value of the prize means that only males who take a high risk in combat can prevail. It also means that most of them will not prevail, and many will pay the ultimate price for trying. This lethal combat among individuals of the same species is perfectly natural; even if muskoxen had a high level of consciousness, they might accept this state of affairs as normal and acceptable. A bull would probably look forward to the rutting season, and at the most opportune stage in his life irrationally charge into the arena, fully expecting to win the prize, but also willing to live with failure and death as an acceptable end.

Of course, muskox bulls do not have a rational brain. They function strictly under the influence of hormones and a set of external stimuli. Their level of consciousness is fairly primitive; they are well adapted to their Arctic environment, but show little sign of problem solving skill or ability to design strategies. They do show an enormous urge to fight, to accept high risks and cope with serious injuries. Perhaps it is overly anthropomorphic, but it is easy to think of these animals as passionately alive in the rutting season, as they first engage in serious combat, and then, when victorious, throw themselves into a mating frenzy. Had high levels of hormones not driven them into a state of reckless passion, would they still be willing to take these risks? A more sober thought is that a bull which avoids serious combat year after year, and eventually dies of old age, will end up with the same zero biological fitness as one that dies as a result of losing in combat. Natural selection favors traits that enhance an individual's contribution to the next generation; if that implies fighting the ultimate fight for access to females, that trait will be selected over any alternate traits.

Outside of the rutting season, bull muskoxen are peaceful animals, which coexist with one another. They have evolved into fighting machines only to gain access to the most important limiting resource they need to succeed in the evolutionary game: females in estrous or about to enter estrous. The willingness to fight a high stakes contest over access to resources, whether they be reproductive potential in the form of females directly, or indirectly in the form of a territory or hunting grounds, is an innate aspect of the nature of the male of many animal species. Of course, this does not mean that fighting is always the best strategy when facing a competitive situation. The typical strategy involves balancing the short-term risk of getting killed in a particular combat, against the longer-term risk of dying with zero fitness due to avoiding combat. All animals have evolved means of sizing up one's opponent, bluffing and bowing out when facing a no-win situation. We humans have very similar innate strategic tendencies, as we have evolved in, and still face, a world of limiting resources.

Back to Anna

Since Anna is not a female muskox, but a very attractive, bright human female, whose rational brain told her that she "should" be happy, why did she set up a situation in which she began to compare her husband of twenty-three years with her seductive professor? She had raised three children with her husband and was committed to her family in a way that gave her the reputation of being a good person. They went to church regularly and were viewed by the community as a "happy family." The simple answer is that she was not as emotionally fulfilled as she had been when her children were in the home and needed more of her attention. She and her husband had begun to take each other for granted years before she returned to the university. She was bored, void of passion, but was unwilling to admit that to herself. She enrolled in this particular class at the same time that the professor was also bored, but he was very aware of what he needed to un-bore himself. They made an unconscious connection through their emotional brains; their amygdalas were starving. Her rational brain didn't stand a chance against the heavy odds. It is easy to stand back and say that she should "forget" her lover-professor, but the more difficult part of her story (and all other similar ones) is that having him in her life made her feel more alive than she had in years. Their relationship allowed her to laugh her laughter and to cry her tears—to feel passion at a new level and to live more fully. It opened an emotional capacity in her that she had forgotten she possessed.

In continuing therapy sessions, she admitted that the last time she had felt such passion was when her daughter was born. She had two sons, whom she loved, but longed for a daughter, although she was afraid to admit this out of the fear that her sons would feel she did not love them. She realized that she had a pattern of denying herself what she really wanted out of her fear of hurting others. She became "the great pretender" to the point that she had lost contact with her dreams and desires, but her emotional brain knew what her rational brain could not admit. I pointed out that she might have known that she needed a passionate romance, but was afraid to admit it out of the fear that she would drive her marriage into the ditch. The only way she could come out of this situation intact was to stop beating herself up for what she called irrational behavior and take an inventory of what had really happened. She began an honest dialogue with her internal wolves. To do this we have to put aside moral judgment and listen to whatever comes. A few weeks into this courageous process, she was ready for a more honest dialogue with her professor-lover and her husband. She did not want to drop the course, but she could see that she could not continue the affair and the course simultaneously. She and the professor spent time getting to know each other throughout the course without continuing their sexual affair. They developed a passionate friendship.

During this time she also decided, with my urging, to invite her husband into therapy and tell him the truth about her feelings and to prepare him for what was beginning to seem inevitable—that she felt she needed a separation. I wanted her to reestablish honest communication with him before she made a decision to throw away her marriage over an emotionally charged desire that might not last. It also seemed fair that he play with a full deck of cards so he could begin to evaluate his feelings and his life. Her most painful dilemma in choosing her course of action was in knowing that she would be hurting someone she loved, regardless of her choice. Could her passion for the professor be worth all the emotional pain she was enduring and causing others? It might be easier to be a cow in the herd and just allow the men to battle it out.

The husband and the professor did not fight as the rutting bulls fought. On the other hand, the husband threatened to kill the professor several times, and in a combat such as the one in which the bulls engaged, he well could have been the victor. What stopped him? After many weeks of therapy, his rational mind finally overruled the passionate hate he initially felt for the professor. He reasoned that he did not want to live with his wife if she were in love with someone else and that he certainly did not want to give up his life, or spend it in prison, if he issued a call to battle.

One of the most surprising, and somewhat unusual, developments in this story is that he requested a therapy session with the professor. I faced a dilemma. Should I set up the session in an open field where they could do battle? How could I insure that blood would not be shed? Anna's fear was that she did not know what to expect. We humans like to think that we will have some control over what might happen. Her husband told her that she was invited to the session as well, which terrified her even more. Trusting my deepest belief that anything can be peacefully resolved if people are willing to participate in an honest discussion, I agreed to monitor the session. My rules were that anyone in the room must come with an open mind and open heart and be able to listen to whoever would be speaking. I knew it was risky, but I also knew it offered the possibility of immense healing. During the two-hour session, each of us gained a deeper understanding and some respect for each person's experience. The last twenty minutes were spent by each person taking responsibility for the pain he or she had caused the others and stating clearly what they would like to have happen. Anna and her husband agreed to file for a legal separation, allowing each other to participate in other relationships.

This kind of situation never plays out without great emotional pain, and often financial strain, but there were no corpses. The professor and Anna are together and their respective ex-spouses are married to other people. Are they happier than they each were previously? They seem to be. Time will tell.

Balancing Our Emotional and Rational Minds

How can we balance passionate emotions with rational thought so that we are able to live passionate lives without destroying our loved ones and our organized communities—without creating a frenzy of human madness? Passion can motivate us to take risks that our rational minds might avoid, or regret having taken. On the other hand, feelings are unmitigated honest reactions and we need to recognize them to stay in touch with the reality of who we are and what we need to experience our aliveness. Lacking the insight and courage to do this, we will become depressed or strangers to ourselves. A dear friend, who works as a hospital counselor, recently shared with me the following story:

An elderly man was sitting beside his dying wife's bedside in the hospital. He had just been told that she would live only a few more minutes, and he requested that their adult children and other family members give him time alone with her. They had been married for almost 70 years. As tears rolled down his cheeks, he gently held her hand, softly telling her how deeply he loved her and how he would miss her. She drew her last breath as he continued to talk to her. He did not move, only nodded to my friend, who had stayed at the door to keep the others from entering. She then opened the door and the entourage of relatives poured into the room. They, although well intended, immediately began consoling each other, trying to console their father/grandfather and organizing what each felt needed to be immediately done. My friend was aware that the grieving husband was becoming agitated by the noise and appeared at a loss as to what to do, so she asked the family to leave the room again. He smiled at her and said, "Thank you. I was about to become a stranger to myself."

This man's need to accept and grieve the death of his wife, while not being disrespectful to his family and of their grief was in great conflict—to the point that he felt he might lose his temper--something he had not often done. Being overcome with passion that is in conflict with our surroundings, with our loved ones, and sometimes even with our image of who we think we are, places us in a difficult dilemma. We have to assess carefully what our greatest need is at the time and in a particular situation.

Our emotional and rational brains meet at that fine red line where they either do battle or agree to compromise. This requires that we diligently stay in touch with our emotional minds, while we remain aware of what we truly value in our lives, thus making decisions that will allow our emotional and rational minds to stay nurtured and healthy. When either aspect of our nature is repressed, disaster usually lurks under the repression. The more we repress our emotional mind, the more apt we are to become like the children with severed amygdalas. The more we repress our rational brain, the more apt we are to behave like monsters. The battles between the two can be fierce, but they each serve a purpose and both are required to maintain our humanity. The only way through emotions is to feel them. There is no real choice, but there is a choice about how to act on them.

Becoming Impassioned

We each possess the ability to live more passionately than most of us tend to do. Let's revisit George. He was not afraid to feel, even when his feelings were negative. He lived on an emotional roller coaster for a few years, but he stayed on board. When he met someone with whom he could share his feelings, he was ready and willing to enjoy passionate experiences. To live passionately, we must be willing to acknowledge our honest emotions. Many of us spend a massive amount of time and energy denying what we really feel in an effort to please others. We become strangers to our emotional selves and lose our authenticity. We pretend to feel what we believe is a proper feeling—a kiss of death to passion.

Anna, unlike George, had become passionless because she was afraid to acknowledge her honest feelings. She was vulnerable to her professor's interest in her because her emotional brain had been on a starvation diet. Opening herself to passion was risky as she risked destroying her marriage, losing her reputation in the community and had no guarantee that her passion for the professor would last beyond the initial attraction. She paid careful attention to her emotions and also engaged her rational abilities in therapy before making her decision. She gave up her fear of discovering her personal truth. To impassion our lives, we must take some risks, give up fear and be honest—listen to that still, small voice that whispers, nags, and reminds us: *This above all: to thine own self be true, And it must follow, as the night the day, Thou canst not then be false to any man.*[11]

George also discovered that one of the keys to unlocking the passion of females was to become a winner—a winner in uniform holding even more power. After completing his stint in the military, he became a successful teacher, an internationally known racing sailor and later, a world champion soaring pilot, all of which gave him the kind of confidence that attracts women, drawn as nature dictates to genetic success. Is it the need to conquer, to win, that ignites passion deep within us, as well as within muskoxen and many other species? We all seem to desire to win at something—even an argument with someone we love. Is our need to win, to be right, greater than our need for peace or for love? Does conflict offer us more passion than love? It has been noted by many therapists, Paddy included, that the most endearing words one can say to another are, "You are right about that." We are often afraid to trust "I love you," out of the fear that we are being manipulated into an obligation. It is time to look deep within ourselves to discover what is ultimately important for us to live a passionate life, while remaining rational and compassionate.

I (Paddy) was fortunate to be present when the announcement was made that the Dalai Lama was to receive the Noble Peace Award. This compassionate man admitted that he had diligently struggled to reach the place within himself that would allow him to forgive the Chinese for taking away the land that rightfully belonged to the Tibetan people. He felt that he had finally achieved this after long conversations between his emotional self and his rational self, but he remained uncertain about inviting the Chinese government in for tea. However, he mused aloud, perhaps the day would come when he could dialogue with them while they drank tea together, when each could understand the reason for the existence of the other.

References:

Q. Osler, Sir William. *Science and Immortality*. In Bartlett, John, *Familiar Quotations*, 13th *Edition*. New York: Little, Brown, and Company, 1955.

1. Tolstoy, Leo. *War and Peace*. Translated by Alexandra Kropotkin. Great Britain: The John C. Winston Company, 1949.

2. Thurman, Howard. In Keen, Sam. *The Passionate Life*. New York: Harper Collins, 1984.

3. Goleman, Daniel. *Emotional Intelligence*. New York: Bantam Books, 1995.

4. Fisher, Helen. *The Anatomy of Love*. New York: W. W. Norton, 1992.

5. Goleman, Daniel. *Emotional Intelligence*. New York: Bantam Books, 1995.

6. Keen, Sam. *The Passionate Life*. New York: Harper Collins, 1984.

7. Porter, Katherine Anne. In Exley, Helen. *Love Quotations*. New York: Gift Books, 1992.

8. Tennov, Dorothy. *Love and Limerance*. New York: Scarborough House, 1999.

9.   Daly, Martin and Wilson, Margo. *Homicide*. New York: Aldine de Gruyter Press, 1988.

10. Laing, R.D. In one of his last keynote speeches during a Canadian Psychotherapy Seminar, Toronto, Canada, 1984.

11. Shakespeare, William. *Macbeth*. In Barlett, John. *Familiar Quotations*. New York: Little, Brown, 1955

Homo Homicides

*People who kill in spite of the inhibitions and penalties
that confront them are people moved by strong passions.
The issues over which people are prepared to kill must surely
be those about which they care most profoundly.*
~ Martin Daly and Margo Wilson

Nathan returned home unannounced a day early from
an out-of-town trip to discover his wife in their bed with
another man. He shot and killed them both.

Deborah logged onto her very wealthy husband's e-mail to
discover that he had been carrying on an affair for several
months. That evening, she sweetly offered him his pre-
dinner cocktail laced with lethal ant poison.

Jeff and his next-door neighbor, Brian, had been at odds for several years over a disputed property boundary. The men frequently exchanged obscenities and engaged in what other neighbors perceived to be harmless threats—until one Sunday morning when Brian began to spray-paint his fence with a noisy generator that kept Jeff from sleeping in on his day off. An enraged Jeff shot and killed Brian.

Nancy, a young single mother, came home from work to discover that the baby sitter, whom she had trusted with her four-year-old son, was sexually abusing the child. She picked up a carving knife and stabbed him to death.

Why We Kill

Although only a small percentage of human beings actually murder another, most of us have entertained at least a fleeting wish that someone would drop dead—or that we could do something to help make that happen. Most of us are stopped by moral or ethical considerations, or by thoughts of forthcoming punishment. Some are not stopped. Some kill in self-defense, or to defend others, especially a child or a relative. Far fewer kill from anger and jealousy. Women usually kill in less violent ways than men, but not always, as in the case of the mother above, who said she grabbed the closest weapon at hand. Millions have killed and have been killed during wars. People of all ages kill themselves to express their rage at what they perceive to be the "unfairness" of life. Some kill for the thrill of expressing their power to destroy another living entity.[1]

A very kind man, who had never intentionally killed anything except mosquitoes and flies, found himself forced into a difficult situation by his family and several neighbors who demanded that he kill a pet goat. Billy, the goat, was hell bent on eating or destroying everything in his field of vision. No other means of trying to change Billy's behavior had worked. With a lump in his throat and pounding heart, he took his recalcitrant goat into the barn and accomplished the task. Much to his surprise, he experienced a brief moment of power, a kind of thrill that he had destroyed the destroyer. He did not expect this reaction. It scared him, but it is not an uncommon reaction. Most of us would agree that he had a justified reason for killing the goat, but he struggled for years with his mixed emotional reactions to the experience. Many men, and even a few women, express similar feelings after performing a life-destroying act. Some of our reasons for killing may be easily justified, such as in war. But many are not, so what is it in human nature that allows us to believe we have reasons to kill, even when the killing may not be justified in the eyes of our society?

It is only in the last half century that scientists, psychologists and philosophers have tackled the question: Why do we become violent enough to kill each other—including, at times, those we love? Is there a connection between passionate love and passionate hate? Killing, or at least thinking of it, seems to be a universal aspect of human nature, as is our revulsion against what we believe to be unjustified killing.

Novels, police records, court procedures and other descriptions of humans killing fellow humans often give reasons or motives for the reported killings. Usually these are proximal reasons. For instance, when a man kills his wife, he may say that he did so because he was extremely angry with her for her infidelity. That would then be recorded as his motive. But we are more interested in the ultimate reasons for murder. We want to know, for example, why we have evolved such violently jealous responses to discovering a spouse's infidelity. Why have we evolved a psyche that is sufficiently passionate about certain situations to make some of us killers? In this chapter, we will explore several different types of violence and propose explanations for such behavior in different animal species—especially our own.

Fish out of Water

Every now and then, we run into a person who in good faith claims that *Homo sapiens* is the only animal that intentionally kills members of its own species. Nothing could be further from the truth. Through technology we may have become the most murderous of species, but killing within the same species is deeply engrained in the animal world. At times it may not be obvious to the casual observer, or it may even be hidden in the complexities of competitive behavior patterns exhibited in nature. However, from lions to doves, murder is part of the struggle for survival.

Over the years I (Dolf) have had several aquarium tanks in which I raised stickleback families, but the first one has remained the one I remember best. My father had explained how the male builds a nest, and then entices a female to lay her eggs there. The male then fertilizes the eggs and instantly becomes a model father, single-handedly bringing up his brood and vigorously defending the little ones from any danger. All that was going to happen in my aquarium! On one of my expeditions to the nearest ditch, I caught a beautiful male stickleback in full breeding coloration. His large iridescent blue eyes shone from a bright red face, and when he moved, his back shimmered with silver and turquoise reflections. I spent many hours sitting by my tank, admiring my new pet, convinced that he was the most beautiful fish in the whole world. He was very active as he explored his new world, often picking up bits of debris and spitting them out again. My parents had a hard time convincing me that both my fish and I needed a good night's sleep.

The next day, my stickleback built a nest. His construction methods appeared random and inefficient, consisting mostly of moving debris from somewhere in the tank to a specific place among some plants. He worked the resulting pile over with his mouth and eventually burrowed through the pile until it became a kind of hut with both a front and back entrance. For the next couple of days he seemed perfectly happy to simply swim around and show off his beauty to anyone watching him.

After a few days of futile hunting in the ditch, I finally caught a pregnant female. She was a smooth, silvery-gray fish with an awkward, bulging belly, which my father guaranteed me contained a clump of eggs. I can still feel the exhilaration and excitement of the moment I dropped the female into the tank. I felt as if I were a divine matchmaker bringing male and female together for the purpose of having a family, and I couldn't wait to see what would happen. Within seconds, the male started to display himself more vigorously than I had seen before; he behaved exuberantly, darting around the female, contorting his body and flaring out his fins as his colors sparkled. Following several bouts of courtship, he started to loop through his covered nest, returning to the female after each loop. During all of this frenetic attentiveness, the female initially appeared to try to escape from him, but eventually, she started to respond more positively to his overtures.

Suddenly, to my utter delight, after the male again swam through his nest, the female also swam into the nest, stopped for a moment and then swam out the other side. Immediately after she left the nest, the male swam back in. He paused for a moment, and, as my father had predicted, the miracle had happened. The eggs were laid and fertilized. The male now busied himself with the nest, while the female, suddenly a slender fish, swam around on the other side of the tank.

I believed that I had witnessed a wonder. Nature had shown me her perfection—perfect love and family life in the world of fish. But reality came crashing down on my head the next morning when I found the female damaged and dead, drifting near the bottom of the tank. When my father explained that the male had killed the female, I was devastated. Although he explained the reason why, it took me several years to fully understand what happened and come to terms with it.

What had happened was perfectly natural stickleback behavior, though the outcome was not. In the ditch or another large body of water, once the male has a nest with eggs, he will defend it vigorously against all potential predators. A non-pregnant female will gladly devour undefended eggs of any species, including her own. The male, therefore, will chase her away from the nest. Once she is sufficiently far away, she becomes harmless. He will leave her alone and return to the nest for guard duty. What went wrong in the tank was that the female could not get far enough away, so the male kept attacking her, eventually killing her. What happens in nature is indeed close to perfect love, as far as stickleback fitness is concerned.

But here's the rub: the perfection of evolved behaviors in nature depends on the environment. For abnormal environments, such as small fish tanks, the male cannot adjust his behavior; when his brain receives a specific set of stimuli, he is driven by hormones to act in a certain fashion. Even when less "hardwired" animals, such as primates, are removed from the environment in which they evolved, they often behave destructively and violently.

Our Own Fish Tank

The stickleback example illustrates an important concept: what a species is and does today has been shaped by natural selection in the distant past, when conditions were very different. Hence, organisms—including human beings—are often maladapted to current conditions. We, *Homo sapiens*, are also trapped in a large aquarium tank. Although our cultural, political and technological world is of our own making, we do not always possess the appropriate innate responses to the challenge our world presents. Natural selection is a very slow process, and we haven't yet evolved all the necessary genetically based adaptations to the conditions we encounter in our current world. Historian Ronald Wright points out in his bestseller, *A Short History of Progress*, that it took us nearly three million years to advance from chipping stone to smelting iron, while we advanced from smelting iron to the hydrogen bomb in only three thousand. This evolutionary gap often puts us in danger and can have disastrous consequences.[2]

Take for instance one of our most beloved pieces of technology, the automobile. Tens of thousands of people get killed each year because we lack the psychological, physiological and behavioral adaptations to keep us safe when driving a car. Our bodies are far too fragile to withstand the sudden impact of hitting a tree or an oncoming vehicle at sixty miles per hour. Yet, our innate competitive psyche makes us want to go faster than the car next to us. We can improve our chances of surviving a crash by engaging our rational brains by using seatbelts and airbags, constructing better-designed vehicles, and avoiding drinking and driving, but many of us fail to behave rationally behind the wheel. Do we choose not to be safe because driving recklessly is a way of expressing a long-ago evolved desire to take risks when pursuing game or an enemy?

Perhaps the most tragic example of the fish-tank phenomenon is our inability to recognize the danger inherent in carrying lethal weapons for our protection. When aggravated by another person's behavior, a natural tendency is to get angry, and anger easily spills over into aggressive behavior. We have evolved to experience anger to defend ourselves. When unarmed, a bout of fisticuffs can cause bloody noses or bruised ribs, but serious injuries are rare. However, a person with a gun may lack the necessary inhibition to use this weapon when angry, often leading to deadly results, as in the case of Jeff and Brian quoted above in the opening section of this chapter. Unless severely enraged, we have a natural inhibition against killing a fellow human being when we are face to face with him, but a gun can be fired at a distance, before that natural inhibition has had a chance to assert itself. This is why it is standard procedure to blindfold those who are condemned to execution by a firing squad, or at least to have them turn their backs. It makes it easier for the members of the firing squad to obey the order to fire.

A Vietnam veteran confided to me (Paddy) in a therapy appointment that when he had once come face to face with a wounded Viet Cong soldier in the jungle, he had not been able to kill him. He said, "The guy was covered in blood, leaning against a tree, but he looked me straight in the eye and asked me for a smoke. I gave it to him. It was as if for that moment we were both normal human beings, not enemies ordered to kill each other. I've thought of him every single day since then and wondered if he lived—and if so, if he ever thought of me. I've also wondered if he remembers me as a coward—or a kind American."

In theory, we should be able to act according to rationally thought-out behavior patterns when handling cars or weapons; however, when our emotions become involved, our choices become more complicated. In practice, adjusting to unexpected situational changes and to emotional reactions often causes us difficulty. It is easy to see why the tiny hard-wired little brain of a male stickleback would not be capable of adjusting its responses to a new and artificial environment. Our higher levels of consciousness and rational intelligence have allowed us to make sophisticated scientific discoveries and rapid technological advances, but our complicated emotional needs have limited our ability to maintain peaceful human relationships—even with those we claim to love and certainly not with those of different religious and political beliefs. Tens of millions of people have been killed in wars with technology specifically designed to kill human beings. Have our fear of differences, our desire to feel superior, and our need to control others combined to overtake what we rationally know—that violence always generates more violence? Why do passion and aggression combine with such intensity that we set aside rationality and take outrageous risks, which often lead to deadly violence? It seems that the more passionately we feel in any situation, the more apt we are not to consider the consequences of our behavior.

The Anatomy of Murder

Recently, during a physiotherapy session for an inflamed Achilles tendon sheath, I (Paddy) was contemplating the profound question of why we kill, when it dawned on me that the physical therapist seemed so gentle, patient, and content to be sitting on his little stool massaging my heel, that I could not imagine he had ever experienced a violent impulse. I asked him outright: Had he ever thought about killing anyone? He let out a snort and laughed so hard I feared he would fall off his stool. "I was in the middle of eight siblings, with three older brothers who terrorized me," he said. "Of course I thought of killing them every day!"

"Oh," I responded. "Anyone recently?"

"Sure. I have teenage kids."

His response was meant as a joke, but it made me wonder. Do most of us entertain, however fleetingly, the thought that we'd like to do away with those who cause us frustration and suffering? The people we love always have the most power to hurt us. They are the ones who push our buttons, so to speak, for they have usually installed them. There is no doubt that my physical therapist adores his family, as most of us do—most of the time. There is also no doubt that when we love, we are vulnerable to the loss of love. When we recognize our vulnerability, we often feel insecure. The insecurity can generate the fear of loss. Fear of losing anything or anyone we deem as precious feeds our desire to protect ourselves and those we love. When we are faced with an imagined or real threat of losing love, self-esteem, property or freedom, our aggressive tendencies are triggered. When fear and aggression are fed, we can become monsters. Let's examine the opening scenarios.

Nathan: Protecting Paternity

Why was Nathan so threatened by his wife having a lover? One reason might have been that Nathan, an Asian medical doctor, had never felt accepted into the family of his American wife, Donna. He resented this and often accused her of secretly harboring a desire to have a lover of her own nationality. He was determined to have children early in the marriage, because he hoped that having children would persuade the family to accept him. Despite her reservations, Donna consented. By the end of their second year of marriage, they had two children. But the tension between them grew. Nathan's medical practice seemed to demand ninety percent of his time, while the other ten percent was spent on the golf course. Donna's resentment of his never being at home and not helping with the children deepened to the point that she asked him for a separation during their fifth year of marriage. He refused. Instead, when he was at home, he pursued her constantly for sex, which caused her great distress because she no longer desired him sexually. "He looks at me as if he could kill me, then expects me to have sex with him," she told me. "He's beginning to frighten me."

Not long after this confession, she decided to attend her 20th year high-school reunion. She invited Nathan to go with her, but he declined and surprised her by volunteering to take the children to an amusement park out of town so that she could be free for the weekend. Donna was deeply grateful, so much so that Nathan became immediately suspicious that she was planning a rendezvous with an old flame. She was not planning a rendezvous; however, it happened that her high-school boyfriend showed up and showered her with attention. She felt a sexual spark for the first time in years. By the end of the weekend, Donna invited her friend to her empty house, where they ended up in the bedroom. Nathan returned early from the amusement park and seeing a strange car parked outside, his suspicions deepened. He asked the children to wait in the car while he went in to tell "Mommy" that they were back. He shot and killed Donna and her lover. When he came back to the car, he told the children there had been an accident and that they were going to their grandmother's until things calmed down. From the grandmother's house, he called 911 and reported to the authorities that there had been a murder-suicide at his home. Following an investigation, Nathan was arrested for the murders, was eventually convicted and remains on death row.

Was Nathan simply jealous and outraged that his wife had sex with another man, or was there more at stake? When we feel we are losing the affection of someone we love, jealousy is a normal reaction. However, experiencing jealousy to the point of murdering two people without any kind of negotiation process, when the penalty could be severe, stretches the definition of normal. We believe that murder under these circumstances is not certifiable socio-pathological behavior, but is at the far range of normal human male behavior, as it is relatively common in all societies. The most persuasive explanation for his violent act lays in the probability that Nathan strongly suspected that his wife was having an affair with another man. His desperate act of murder was a response to the most serious possible threat to a man's Darwinian fitness. His own future reproductive output was in jeopardy. If his wife produced a child by the other man, he could very well end up raising and supporting that child instead of another child of his own.

Men are more likely to kill for being cuckolded than women. Of course, most men under such conditions do not consciously think in terms of lost reproductive output, and hence, lost biological fitness. But no trait comes closer to what natural selection acts on than reproductive output. A competitor who threatens to directly lower a man's reproductive output will be perceived as a mortal enemy. As Donna's lover belatedly discovered, cuckoldry is a risky business. Why did he take this risk? Again, it is unlikely that he was consciously hoping to extend his fitness by getting Donna pregnant; they may well have been having safe sex. But his passionate desire for sexual contact with Donna was part of his nature.

That desire, in turn, was a direct result of natural selection because sexual intercourse has always tended, sooner or later, to lead to pregnancy. Why did Donna take the risk? Enticing the old flame into her bed is an evolved behavior that is based on the possible increase in reproductive value of a woman seducing a man she considers of greater value or status than her mate. She had come to think of Nathan as uncaring and demanding, while she perceived her old flame to be sensitive and loving. These traits would probably make him a more reliable spouse and father. We must assume that over many past generations, seductions, cuckoldry and jealous violence must have paid off in an evolutionary sense, at least on the average. For some, it has been a very bad risk.

Deborah: Keeping it in the Family

Now, we turn to Deborah, who poisoned her wealthy husband after discovering that he was involved with another woman. She had been suspicious that he was having an affair and had even boasted to friends that she hoped as much, as it would give her a reason to leave him and take half of his money. Several of her friends had suspected that she married him for his money in the first place, as she seemed more passionate about her wealth and status than about her husband. One friend had warned her that if she married him for wealth, she would pay for every penny of it. Currently, while still occupying a position on death row, she may well remember that warning.

Although we know that the number of men who kill their spouses far outnumber the number of women who do so, it is interesting that the warden of the prison, where Deborah is being held, reports that Deborah receives phone calls every day from other women who want to know, "Was it worth it?"

It is easy to surmise that Deborah's primary motive for marrying and for killing her husband was greed, and that discovering his involvement with another woman just gave her a pretext for murder. This case becomes more interesting from the standpoint of evolution, as Deborah had three children from a previous marriage. Could her ultimate motive have been to ensure that her children inherited some of the wealth accumulated in this marriage and to prevent that wealth from being diminished by a portion going to another woman and her children?

In modern human society, one does not only inherit good or bad genes, but also wealth and social status, which in the next generation can be translated into increased biological fitness. In evolutionary terms, Deborah's fitness was represented by her three children and would eventually be transferred to her grandchildren. Her husband's wealth, her proximal motive for his murder, would undoubtedly increase the opportunity for her children to have more affluent families. We doubt that as Deborah murdered her husband to gain control over his wealth, she was consciously thinking about the welfare of her future grandchildren. We know, however, that natural selection has made us unconsciously desire to gain wealth and transfer it to our children or other "blood relatives." This is especially true for older women, who can no longer contribute directly to their biological fitness through childbearing. They tend to contribute through wealth or by providing support and care to the nepotistic component of their biological fitness.

Jeff and Brian: When Winners are Losers

What about Jeff and Brian, the two warring neighbors? They hated each other, but almost seemed to enjoy sparring for power via insults and attempts to turn other neighbors against the "enemy." The fact that Brian had put up a fence on what Jeff considered his property line was the stimulus for their animosity. Jeff believed the fence to be located on the easement between their properties, where he had once requested to put in a driveway. His request had been denied by the municipal authorities, so he was infuriated over Brian's fence and let it be known throughout the neighborhood that his neighbor must have paid off the survey team. This attempt to harm his reputation equally infuriated Brian. Then, on the fateful Sunday morning, while Brian was painting the contested fence, with a loud compressor driving the sprayer, Jeff called out the window for his neighbor, to "cut out the commotion." Brian, who was wearing earplugs at the time, did not respond. Jeff shot and killed him.

During the trial, Jeff and his family convinced the jury that he had never meant to actually shoot Brian, but only to shoot the gun close enough to get his attention. Since Brian was moving quickly down the fence with the sprayer, Jeff's story had some degree of credibility. He was not convicted of first-degree, pre-meditated murder, but of third-degree murder, and sentenced to fifteen years in prison. His family was so humiliated that his young son refused to go to school, reporting that the other children called him mean names. His wife sold their home, moved with her son to another state, and has recently filed for a divorce. Jeff not only lost his status, his ultimate motive for continuing this stupid feud, but everything he loved.

Many high testosterone males who impulsively kill over property, status or money express regret over having killed their opponent. Yet, in the moment they are so carried away by the need to be "the winner" that they do not consider the consequences of their violence.

In each of the above cases, there is a component of the "fish tank" concept. Had Nathan not owned a gun, he might have beaten up Donna and her lover, but no one would have died. In early human societies, a man's wealth, upon his death, would always go to blood relatives, never to his wife. So the Deborah of that era would not have been tempted to murder her rich husband. Until relatively recent times, arguments of the Jeff and Brian type were often resolved with pistols or swords, and the party that fired the lethal shot was rarely penalized. Organized duels were common until well into the nineteenth century and were not considered murder. But we must address the problem that Jeff and Brian were not fighting over a woman or wealth. Several generations ago, in most societies young men were prevented from marrying or from having casual sex with women until they had acquired sufficient status within their community. Gaining such status was usually a serious, competitive business. In contemporary society, men like Jeff and Brian might have gained more social status had they availed themselves of counseling and/or mediation instead of responding to their passionately competitive instincts.

Nancy: Defending Her Child

Then, there is the kind of murder that pulls at our heartstrings, the case of Nancy, a mother who killed to defend her child. Most of us can easily forgive a mother for allowing her passionate love for her child to over-rule her rationality. In this case, the jury did forgive her. Nancy has become a heroine of sorts for preventing the young man from abusing her child. In evolutionary terms, this is the simplest case; the mother's Darwinian fitness is expressed in the child and possible future children. Most women will protect their offspring fiercely. Whereas Deborah's grab for her husband's wealth could have enhanced her fitness indirectly, and Donna's promiscuity added somewhat to her potential reproductive output by possibly having a child with her lover, Nancy protected her already realized fitness directly. It is not surprising that the jury let her off. Society obviously sees the difference between the above cases. Deborah and Donna took the initiative to seriously harm their husbands, and paid the ultimate price, while Nancy killed, in an evolutionary sense, in self-defense.

In all of these cases, a human being's overwhelming passion is combined with aggression to destroy another human being. Love is also involved—love of progeny, wealth, status, or of someone we wish to protect. Let's take a deeper look at how love and passion play out in risk-taking and killing.

The Passion Trap

Fairy tales in numerous cultures focus on the theme of a young man who takes severe risks to defeat some evil before he can win the hand of a highly desirable young woman, usually a princess. These tales usually allude to other pursuers who have failed, but the hero, driven by romantic passion, eventually prevails. He invariably ends up living happily ever after, with a beautiful wife, high social status, wealth to spare, and presumably a large family. Our rational advice to such an impulsive young man would be to think twice, to consider his minute chance of success, and let another fool face the dragon or the evil king. Yet we continue to delight in these fairytales. They were read to us and we read them to our children and grandchildren. What is the explanation for the enduring appeal of this irrational, passionate risk-taking in the tales of so many cultures?

Men from our hunter-gatherer era onward have had to prove themselves by facing and surviving dangerous situations before they could gain sexual favors from women or acquire wives. Only those men who took risks and prevailed would be able to contribute their genes to the next generation. But remember, to throw yourself into battle, to tackle a bear with a spear, or to court the daughter of a fiercely protective father, one must throw all rational caution to the wind and let passion carry you onward. Again, those who were passionate about risk-taking, despite their high failure rate, must have left more offspring than those who cautiously waited for a safe path to success. An explanation could be that the heroic winner casts his genes liberally into the future, while the losers, or the ones who stay at home, do rather poorly in this department. We respond to these tales because they express some of our deepest evolutionary drives. But, there is a further complication. These stories encourage most of us to dream of being a dragon-slaying hero, to unrealistically assume success, and to revel in the expected high status (and hence increased fitness.) However, these rewards are not the direct result of destroying the dragon or the evil king. There would be no reward at all, were it not granted by society, which is only too willing to encourage—and often sacrifice—young men to try to protect Society from undesirable trouble-causers.

The young men of our era, as described by social psychologists, drive cars too fast, compete for girls at parties, experiment with drugs, and get themselves into fights—fights that sometimes leave somebody dead.[3] Even though, in contemporary Western society, killing your competitor can get you life in prison rather than the hand of the princess, and the wealthy man's daughter you seduced gets an abortion, your risky behavior has not benefited your fitness one bit. But it is still de rigueur not only to be passionate, but also to display one's passion under socially acceptable circumstances. We praise those who are passionately in love when legitimately courting. We demand passionate patriotism from all citizens during times of national crisis. Most cultures, including our own, worship successful risk-takers, winners who stand out for having the courage to try again and again. They are the names we recognize, and that recognition is a validation that an individual can make a difference. It gives each of us hope that we can make a difference: that our lives count for something. Deep inside, all of us seem to harbor a desire to be a hero or heroine. We yearn for someone's applause, recognition and love. If winning that applause requires that we throw caution to the wind, many are willing to do so.

Natural selection has shaped us into aggressive competitors when it comes to gaining status, wealth and sex, because these aspects of our lives are so strongly related to the continued survival of our genes. Society condones and encourages our passionately competitive behavior, but only as long as we do so within the law. Whereas most of us recognize the limits set by society and are able to live within them, some lack that control. Given our society's adulation of successful risk-takers, combined with our powerful genetic heritage, it should not surprise us that at the extreme edges of the normal range of human variation there are some who overstep the line.

Pathological Violence

Beyond the cases of lethal violence already discussed, those that can be explained within the combination of societal stresses and our evolutionary past, there are some rare cases that do not fit into any rational, or even explicable to date, pattern of human behavior. We have to assume that the perpetrators of repetitive lethal violence have a socio-psychopathic disorder. Clinically speaking, a socio-psychopath is characterized by repetitive, flagrant and remorseless violation of the rights of others and the rules of society. They are described as lacking a conscience and seem unable to feel empathy towards others; therefore, they are unable to love. Bottom line, they seek impulsive self-gratification and place no value on the lives of others. This can certainly be said of serial killers, such as "BTK" (Bound, Tortured, Killed), an animal-control officer, recently captured, after three decades of ghastly murders. This man was married, a father of two grown children and was active in his church and community. He definitely fits the description of a vicious wolf in sheep's clothing—and of a socio-psychopath.

Another particularly brutal example is Paul Bernardo, who as a young man in his twenties, relinquished his career as an accountant with a large financial institution and started a life of crime. He had grown up in a dysfunctional family, with a father who sexually assaulted his own daughter. However, as Paul matured, he became known to people who met him as charming, highly intelligent, ambitious, well groomed, handsome and very successful in attracting women. Despite his easy access to willing women, he started to accost young women, and soon after, he began brutally raping his victims in a cruel and degrading manner. His intelligence, his day-time gentlemanly appearance and a dose of luck allowed him to continue this behavior for a couple of years, raping at least fourteen mostly teenaged women.

Massive efforts were put into trying to capture him, but the police were unable to corner him. To the public he became known as the "Scarborough Rapist." After he moved to a different city, he extended his life of crime, earning a good income from managing smuggled contraband and began to seduce or kidnap teenagers, forcing degrading sexual acts on them. After murdering three of his victims, he was finally caught. In custody, he has shown no remorse, no sense of guilt, and considers himself a victim of circumstances.

The behavior of these men (It is extremely rare for females to be diagnosed as socio-psychopathic.) is certifiably a socio-psychopathic disorder, with a strong emphasis on sexual sadism. Although bizarre forms of sexual gratification are more often than not an aspect of the socio-psychopath's behavior, there are cases that involve torture, killing and even cannibalism, without sexual involvement. There is a growing amount of evidence that this rare behavior is an inherited syndrome, which diminishes the possibility of curing the behavior through acceptable means available in most societies. Usually, socio-psychopaths end up in maximum-security prisons, housed in the wards specified for the criminally insane. Some end up on death row and are executed.

There is also evidence that severe trauma and abuse can break down the psyche and destroy the conscience to the point that socio-psychopathological behavior develops as a result of this history. This theory will be further explored in following chapters in relation to tyrannical systems of government and the resulting vanity wars in which thousands are forced to kill and are killed. War demands socio-psychopathic behavior. It is worthwhile to consider whether the traumas of war might even increase the incidences of socio-psychopathic disorders.

Cases of socio-pathological violence have been reported since the beginning of our recorded psychological history. Many genetic and social theories have tried to explain its origin, yet none seem to apply across the board. We have to question what aspect of this abhorrent disorder prevents it from being eliminated through natural selection.

One plausible, but as yet not generally accepted explanation, sees male psychopathic behavior as an extreme case of the spectrum of the male reproductive strategy in which the male maximizes the number of progeny without allocating any time or resources to the care of his mate(s) and offspring.[4] It follows from a more elaborate version of natural selection called frequency dependent selection, which favors a trait when its frequency in a population is low, but disfavors it when the frequency is high. This leads to a stable equilibrium at some frequency, but can get a selective boost under conditions of societal breakdown, such as prolonged war or anarchy. An example is found is several species of salmon, in which mature males come in two sizes. The predominant large males construct nests, which they defend, and in which the female deposits their eggs. The smaller males wait near such defended nests, and as soon as the females deposits the eggs, the smaller male darts in and squirts some sperm into the nest before the larger male can take action. Natural selection favors this "sneaker" strategy only as long as it is rare, because its success is contingent upon the larger males' protecting the nests.

Hawks and Doves

The males of some species, such as muskoxen and lions, are serious risk-takers, who fight deadly duels for domination of a group of females. Humans are dealing with a more complicated social structure, which means that individuals are to a degree controlled by society as a whole. This usually implies certain limits on violence, with the rare exceptions of those who exhibit socio-psychopathic disorders; they become self-absorbed to the extent that they are not concerned with the feelings of society. In general, we can say that most humans have evolved to desire approval and acceptance by society or by members of their tribes and families. Some, willing to prove themselves with serious risk-taking behavior in order to gain acceptance and approval, could be called the Tamino type. Tamino is Mozart's hero in the opera, The Magic Flute, who, upon going through the rites to be accepted into a secret order, is asked by the speaker for the order, "Stranger, what do you seek from us?" Tamino replies, "Friendship and love."

The speaker then asks, "Are you prepared, even if it costs you your life?"

Tamino replies, "I am."

Some societies do require that young men prove that they are willing to risk their lives for friendship and love. For instance, among the Pokomo people who live along the Tana River in Kenya, the traditional manner in which a hopeful young man must prove his worthiness to the parents of his beloved is to partake in a dangerous, ceremonial lion hunt. He not only has to prove his courage, he also has to prove that he fully trusts his best friends.

While living in Africa and traveling along this river, my wife and I (Dolf) met a young Canadian physician who had spent a large part of the day treating two men who were badly mauled by a lion. The ceremonial nuptial hunt had involved three men—the young groom-to-be and his two best friends. They set out to find a lion, the two friends armed with spears and positioned on either side of the unarmed lover, who merely had his right arm heavily swathed in thick cloth. The intention was to harass the lion until it attacked. In theory, the lion will always attack the man in the middle, who will push his wrapped arm into the lion's maw as he falls backwards with the lion on top of him. Immediately, the two friends will then kill the lion with their spears.

The fact is that enraged animals don't always behave according to human wishes. According to the doctor, the lion had jumped one of the friends instead of the future groom. Meanwhile, the other friend had broken his spear by thrusting it the wrong way. Upon sensing a new, more urgent foe, the lion promptly attacked the man who had broken his spear. This desperate situation was resolved by the unarmed middleman, who picked up the spear of the first victim, and drove it home, instantly killing the lion. The good doctor was able to save both victims, and the groom-to-be got the nod from his future in-laws. Should it be surprising to us that young men in our society race their cars on the highway, and partake in other risky behavior to establish status or impress girlfriends?

Some other species are much more cautious than humans, engaging in what we might label ceremonial or symbolic fights to determine the stronger contestant. At a certain point in a conflict, the weaker competitor will bow out and try his luck elsewhere. Significantly, this less lethal approach is typical of animals that are strongly monogamous. Among the Arctic fox, for instance, there is very little direct conflict between males over access to females. Well before the breeding season, the male and female fox form a pair bond and stay together for the entire season. Monogamy in itself does not necessarily insure a peaceful life. To raise a large litter of healthy pups, the foxes have to claim and defend an adequate territory with sufficient resources, which often brings them into direct conflict with neighboring pairs. Although direct lethal aggression is rare, the more dominant pairs will chase weaker ones away, thereby depriving them of the opportunity to breed and raise a litter successfully.

At this point we might ask why humans are not like foxes? Are we monogamous? This question requires our digging a bit deeper into the relationship between mating systems and the structure of human society.

Life Strategies

For each species, natural selection has caused the gradual evolution of a specific life strategy. This is a well-integrated package of several behavioral categories, such as foraging, courtship, breeding, raising young, defense of territories and other social interactions. Such a species-specific strategy may or may not involve lethal conflict among members of the species. In the muskox, a highly polygynous breeding strategy has evolved, resulting in serious lethal aggression among males for the acquisition of a female herd. (We use the term polygynous because it consists of two Greek roots, polu [many] and gune [woman], instead of the more general polygamous, which means the practice of having two or more spouses at the same time.) By contrast, as we have seen in the monogamous Arctic fox population, there is no overt aggression among males for access to females. The differing levels of aggression in different species are always related to other parts of the life strategies of the species and to the ecological conditions. Natural selection acts on entire individuals, not on single traits or genes. It is also true that animals with life strategies that involve lethal conflict have evolved their strategies partly because of the fitness value of the conflict trait.

The Arctic fox depends on a reliable supply of lemmings in order to raise a family of kits. Lemmings are fairly sedentary small rodents, which can be quite abundant, but it takes considerable skill to catch them. This ecological reality for the fox has been the basis of natural selection resulting in the life strategy of the fox including the defense of a territory with an adequate food supply. It also favors a monogamous breeding system, as it takes two cooperating parents to successfully raise their brood. On the other hand, Antarctic fur seals also have a life strategy that is strongly influenced by two critical biological factors. They depend on highly mobile schools of krill (a large shrimp-like organism), which cannot be defended, and the females cannot give birth in the water; they must drop their pups on a beach. In response, the males have evolved a strategy to wait on the beach for the females to come ashore. They fight vigorously among themselves for a good patch of the beach, where they attract a small herd of females. After the pups are born, the females nurse them between long absences to feed on krill in the ocean, while the males guard the pups. On such beaches, there is constant violence among the males until the females, which enter into heat shortly after giving birth, have been bred.

Our life strategies also include a wide range of complex social behaviors and dependence on others. This interdependence means that lethal violence within any organized group is relatively low, though not absent. We are vastly more likely to kill beyond the group, in warfare for example, than within the group. Male – male competition for status within the group is often externalized beyond the group, allowing men to gain status as heroes in warfare or other situations that compel us to subdue a perceived common enemy.

Till Death Us Do Part?

Those of us sharing the planet in this 21st century are the result of many generations of natural selection. Our chimpanzee-like ancestors of some six million years ago were already living under conditions where a male could increase his fitness considerably by mating and producing offspring with more than one female. Over hundreds of thousands of generations, males who competitively succeeded at fathering offspring with more than one female drove natural selection to make us the moderately polygynous species we are today.

Many horticultural tribes, such as the Xavante Indians of Brazil, practice polygyny. The most successful men of this tribe have over twenty children, while the most productive women have only eight. Other hierarchically structured societies, such as the pre-colonial African kingdoms of Ashanti or Baganda, have greater variance in reproductive success among males. In societies where monogamy is the legally enforced reproductive relationship, some extra-marital reproduction still takes place. More importantly, due to the individual human male's longer period of fertility, the potential ratio of reproductively active women to men is much smaller than one on one. A significant percentage of young women become second wives for older men. Both extramarital sex and sequential marriage mean that certain males will have many more offspring than others. One of the cultural icons of Western civilization, Johann Sebastian Bach, fathered twenty children by two wives in sequence.

We are diverging into mating practices in this chapter because polygyny, the de facto mating practice of our society, breeds violence. The essence of polygyny lies in the simple fact that women are limited in the number of babies they can produce, while men are much less so. A woman cannot increase her fertility by mating with more men, while a man can increase his fertility considerably by mating with multiple women. This means that men can increase their fitness by fighting other men, taking risks, raping unprotected women, and by enticing women into their domain with promises of wealth and protection. This creates a situation in which men cannot acquire a mate or mates without competition. Nathan and Donna's lover were essentially fighting over a woman. Jeff and Brian were indirectly and subconsciously doing the same thing; social status gives a man a distinct advantage in this potentially lethal mating game.

That fighting and risk-taking is much more a man's than a woman's strategy has frequently been explained as a function of men's generally larger size and greater strength. In fact, the opposite is true. Males of many species, including the human species, have evolved a larger size and greater strength because they do most of the fighting. Besides the convincing evolutionary arguments explaining this situation, there is some intriguing evidence. Men have higher metabolic rates, which gives them an advantage in the competitive world, but also places a strain on their immune systems and other more sophisticated internal life regulating systems. This imbalance contributes to the lower life expectancy of men as compared to women.

Women also compete among themselves, but not for just a mate. Women compete for the much smaller, but not insignificant, benefits that come with males of high quality, rank or wealth. Women can have the same number of children with a poor and uneducated husband as with a rich, educated one. Their children will probably be healthier and eventually of higher status if they are fathered by a man of high status, but the advantage is relatively small. In other words, the difference between losing and winning in the mating game is much less for women than it is for men. This implies that women, having less at stake, will take fewer risks. They tend to compete with strategy, manipulation and subterfuge, rather than with violence. Our polygynous mating practice also explains why men do the courting, and women do the choosing. As the saying goes, a man chases a woman until she catches him.

Women rarely fight directly with one another, except over sparse resources when their offspring face starvation. Females will also fight males for the protection of their young. Polar bear sows will fight a male when he attempts to snatch her cub. In such cases, the male tends to give up, despite his greater strength, because for him the fight is merely over a meal, while for the sow, the stakes are an entire year's fitness output. Similarly, Nancy aggressively protected her child when the babysitter molested him. While Nancy was unlikely aware that her child represented her biological fitness, this is at the root of a woman's instinctive protective behavior toward her child.

Recent research has unexpectedly revealed that we humans have something important in common with songbirds. Biologists had always assumed that songbirds were monogamous, faithful pairs sharing the responsibilities of bringing up their nest full of chicks. They were like role models for us less virtuous humans. But DNA analysis of song-bird parents, nestlings and neighbors proved otherwise. In some nests, a remarkable percentage of nestlings were fathered by neighboring males. Follow-up behavioral studies have since revealed that the females are the ones who cheat on their mates. They quietly leave their territory and elicit copulations from neighboring males. In most cases, the cheating females are paired with relatively low-status males and present themselves to higher-status neighbors.

Other songbird species are blatantly polygynous. In these species, the male tries to attract two or more females into his territory, while the females have to decide which is a better deal, joining an already paired male as his Number Two, or joining a lower-quality, as yet unattached male as his Number One. Meanwhile, the males engage in a lot of fighting along the territorial boundaries, trying to enlarge their fief and showing any female who may be watching or listening what impressive he-males they are. Does any of this sound familiar?

Living Together; Controlling Violence

Animals of species that have evolved a social life strategy of living in close, interactive proximity to one another have the advantage of more protection from predators, but they also have less individual freedoms. Organized groups set limits on individual behavior in order to maintain stability for the group. This certainly applies to the ways we have structured most human societies. Among mammals, the muskox has evolved a fairly complex herd structure for protection against predators, which allows for lethal aggression between males, but this violence is restricted to the rutting season. Embedded into this herd structure is even some evidence of empathy and altruism.

I (Dolf) once witnessed the birth of a muskox calf, which made me think that even these simple grazers have evolved a bit beyond selfish opportunism. My colleagues and I watched the herd from the top of a hill, unnoticed by the animals. It was obvious that the lead animals wanted to move on, but after a few steps, they would turn around and watch a single cow that was lingering behind. Suddenly, the lone cow gave birth and started to lick the calf, which was lying on the grass. After a while, the cow nudged the calf, as it tried to stand up. Finally it succeeded and staggered around a bit on shaking legs. Immediately, the herd started to move on, but very slowly. It was obvious that the other animals were responding in a way that ensured that the cow and her newly born calf would not be left behind. These extremes in muskox behavior, from highly aggressive to caring and considerate, are part of our human legacy as well.

Social interactive behaviors and the structure of animal societies reach their peak of complexity in the primates, of which we humans are the most advanced. As the best example of a non-human social structure, we will focus on the chimpanzee, because this species is one of our closest relatives and has been studied more intensively than any other animal living in complex communities. In the wild, chimpanzees live in very tight communities, with clear hierarchical social structures. Mature males have a hierarchical arrangement based on a combination of physical strength, age, personality and social support. Females and young are always of lower status than any of the mature males. The female hierarchy is a complex matrilineal affair, which carries status in the troop from mother to offspring, even to males as they grow up. The entire social structure is not very stable because altercations ranging from minor scraps to major fights—with potentially lethal consequences—tend to result in hierarchy rank-order changes. The alpha male maintains his position by intimidating displays of power and occasionally by fighting off contenders.

Frans de Waal has studied and described a further complication, the establishment of alliances of two or more animals that can significantly affect the hierarchy, even to the point of demoting or killing an alpha male, or alternatively protecting an aging alpha male from being demoted. While female alliances have also been reported, they are not aimed at acquiring a higher position in the hierarchy; rather, their aim seems to be to undercut the despotism of the alpha male through group intimidation. In general, the social hierarchy is maintained by violence and threats of violence. However, the dominant role of the alpha male and female alliances can also lower violence within the community through intimidation.[5]

Beyond the Pale

Whereas there is good evidence that chimpanzee troops have mechanisms in place that lower intra-group violence, it is a different story for between-group aggression. Between-group interactions usually involve no more than loud intimidating shouting matches, but when two communities actually come into physical contact, lethal aggression is normal. Sometimes, intentional forays are made into the other community's area for the purpose of attacking and killing a member of the other community. During times of food shortages, communities have been observed taking over the entire territory of a neighboring one, and killing the neighboring males, one by one. A single male chimpanzee is never welcome to join another community. In fact, when found alone by a group of neighboring males, he will be attacked and killed. Females, on the other hand, have been reported to leave their home community and be accepted by another.

The distinct difference in frequency and intensity of violence within a community and between-communities is of considerable importance and deserves explanation. Since the status of a male chimpanzee is closely correlated with his opportunity to copulate and hence, with his reproductive success, the frequent but rarely lethal violence resulting from a male's attempts to climb the hierarchical ladder is easily understood. It is not that different from what we encountered in the muskox, but tempered by the male's need to maintain a safe position within a more or less cohesive social unit. Similarly, the need for a good-sized territory with adequate resources can explain the concomitant need for between-community warfare.

But why are individual males willing to partake in risky forays into enemy territory for the purpose of finding and killing closely related individuals? We have to assume that the male chimpanzee has an emotional drive to go after and kill, at considerable risk to himself, what he considers an alien competitor, a stranger who is a deadly threat. A male of another community is the dangerous "other." The imaginary line drawn in the sand between the territories of two adjoining communities is like the pale of a medieval village, and he who lives beyond the pale is evil and dangerous. Once an individual sees another as alien and threateningly evil, the natural inhibition against killing falls away.

We still have much to learn about human evolution pertaining to our social and cultural development. While physical anthropology and archeology are starting to shine glimmers of light into this void, some critical questions about our evolution from ape to man can still be tackled only by theory and speculation. Social control over behavior, of which we see the beginnings in chimpanzee society, must have strengthened step by step to reach the levels we have observed in Stone Age societies in places like New Guinea and the Amazon region of South America.

Some of these societies remained nearly completely isolated from outside influences until well into the twentieth century. Such primitive societies enforce strict rules that limit violence within their groups or village communities. They have successfully externalized violence beyond the pale, so to speak. Because defense of hunting grounds and land is essential, an unprovoked surprise attack aimed at a potentially threatening neighboring clan or village is a more successful strategy than waiting for the neighbors to strike first. The seeds of human warfare were sown millions of years ago in a primate ancestor of ours. From then up to now, neither natural selection, nor treaty negotiations have succeeded in replacing warfare as the preferred solution to survival in a competitive world. Why?

References:

Q. Daly, Martin and Wilson, Margo. *Homicide*. New York: Aldine de Gruyter Press, 1988.

1. Ibid.

2. Wright, Ronald. *A Short History of Progress*. Toronto, Canada: House of Anansi, 2004.

3. Keen, Sam. *The Passionate Stages of Life*. New York: Harper Collins, 1984.

4. Frank, Robert H. *The Strategic Role of the Emotions*. New York: W. W. Norton. 1988.

5. de Waal, Frans. *Good Natured: The Origins of Right and Wrong in Humans and Other Animals*. Cambridge, Massachusetts: Harvard University Press, 1996.

CHAPTER

Levels of Loyalty

Man did not enter into society to become worse than he was
before, not to have fewer rights than he had before, but to
have those rights better secured.
~ Thomas Paine (1791)

As we begin to answer the critical question of why a thinking society of humans would continue to compete to the death with another society for resources rather than explore all methods of cooperative behavior, we must examine the challenges posed by our dual disposition of competition and cooperation. Since the beginning of time, these opposing tendencies have been reinforced through our survival skills, and each serves an important purpose. Most of us have great difficulty repressing our strong natural urge to compete. It feels good to win, to be number one—to take home the prize. Yet, if we are to survive as a civilized species, we must begin to understand that the prizes for "winning wars" are running out. We must do more than dream of peace, we must commit to developing more cooperative ways of balancing our needs and resources so that peace can prevail.

One of the most impressive displays of cooperative social behavior in a non-human animal is "bubble-net fishing" by humpback whales. On several occasions, I (Dolf) have stood in awe on the deck of the Lindblad expedition ship *Sea Lion* in the Alaskan inside passage and watched a pod of eight to twelve of these thirty-ton leviathans perform a feat of underwater cooperation that boggles the mind. When the pod locates a large school of herring or another suitable prey, the whales dive deep under the herring school and swim around in a tightening circle, blowing air into the water. This has the effect of trapping the small fishes in a water column surrounded by rising bubbles. Then, the whales congregate at the bottom of the column and suddenly shoot upwards, opening their mouths wide as they encounter the dense school of confused herring. What we could observe from the ship was a ring of bubbles on the surface, surrounding the rippling water full of fish, followed by a dense pack of open-mouthed whales crashing through the surface in an explosion of water and small wriggling fishes. As the humpbacks sink back under the frothing surface, more water spouts in all directions as the whales force the water from inside their enormous mouth cavities through the baleen sieves, which keep the herring from escaping. On some days, when the herring schools are dense, the humpbacks will repeat this incredible performance over and over again.

What is so amazing about this hunting method is that it is not a hard-wired intrinsic behavior. All evidence suggests that it is a learned activity. The humpback whales of Alaska spend only the summers in these resource-rich waters; come fall, they migrate to the warm tropical waters around Hawaii, where they go through their courtship and mating rituals shortly after the cows give birth to their calves. The humpback is a highly social species that uses a song to communicate with other humpbacks over very long distances. Single individuals, cows with calves and pods of various sizes seem to be in vocal contact with others at all times. Their cooperative fishing and the males' courtship songs are wonderful examples of sophisticated animal culture. The courtship songs, like the songs of our pop culture, differ from year to year and from region to region. New songs rapidly spread, with all males trying to out-croon one another.

In these whales, as in every species, conflicts between individuals or pods are inevitable. Yet, while males in courtship do a lot of shoving and pushing, there is little evidence of overt violence. On the feeding grounds, instead of competing for the best patches, cooperative use of the resources seems to be the preferred strategy. When resources are scarce, the animals spread out or migrate. On the whole, humpbacks are extremely peaceful creatures. We might benefit from finding out more about the structure of humpback society and the selective pressures that have made these whales so collaboratively peaceful. The creatures we have hunted as a mere source of blubber and baleen may someday teach us a lesson or two.

Hockey Player's Dilemma

We humans certainly face the dilemma of how to deal with our conflicts over resources, but, as we are painfully aware, we are usually far less peaceful in resolving them. Ideally, we must balance our rights and needs with those of others within the framework of our society. This implies integrating several levels of responsibility, beginning with the individual responsibility to live up to personal goals, followed closely by our responsibilities to family and friends, then on to our commitments to the society with which we identify most closely, various higher levels of government, and eventually to humankind as a whole. What makes this mandate both interesting and complicated is that the various levels of responsibility are partly in concert and partly in conflict with one another. Boundaries between the levels can become impossibly vague, as we struggle to make decisions about which behaviors are appropriate and beneficial for each level. Because this ambivalence is so clearly illustrated in the simplified world of competing sport teams, we call this conflict the "hockey player's dilemma."

If you are a member of a professional hockey team, you naturally want to be the star of the team, as that status could gain you an extra million dollars a year. But you also want the team to win its games. This is the first stage of the dilemma. To be the star, you have to score more goals than your fellow teammates. On the other hand, to win games, you have to be a team player who will generously pass the puck when a team-mate is in a better scoring position. The second stage of the dilemma coalesces around one's level of aggression towards the opposing team. The more you can intimidate and rough up members of the other team, the better—as long as you don't violate the rules and saddle your team with penalties. While these can be difficult dilemmas for athletes, playing the optimum strategy in the game of life is vastly more challenging.

Our First Dilemma: Self versus the Common Good

In previous chapters, we have already seen a glimmer of the concept "for the common good" in a few individuals of the chimpanzee society. Among social animals, cooperative behavior, such as group hunting, building shelters and joint defense activities, is easily explained through natural selection. To survive and thrive, members of a society must help each other. However, making a sacrifice for one or several members of one's society is a different matter. How do we explain those members of a chimpanzee community who are willing to risk insult or injury by trying to separate two fighting males, or by trying to get them to reconcile after the fight? When the alpha male takes on this role, it is understandable, since he incurs little risk and reaps the benefits of reconfirming his top status, and thereby gaining greater access to females in heat.

However, primatologist Frans de Waal describes how some females also take on the mediation role—with much greater risk of injury. When two males have had a fight, they often remain in a huff, angry and quite capable of either resuming the fight or of taking their frustration out on other, lower-ranking members of the community. In such a situation, a female might take the initiative to calm the males and bring about the reconciliation of the two hotheads. Her typical strategy is to approach one male and kiss or touch him gently. Then, once he seems somewhat calmer, she leads him towards the other male, keeping herself between the two adversaries. Soon, they start to groom her. When the tension has further abated, she gently withdraws, leaving the males grooming one another. In an even riskier maneuver, females occasionally try to calm down two males who are gearing up for a fight.[1] Why would females put their bodies on the line this way?

100

Natural selection acts on individuals, favoring those whose life strategies enhance their survival and reproductive output. *Some individuals do this more effectively than others of the same species.* This is important because a female chimpanzee who plays the role of mediator between two riled-up males takes an increased risk of getting injured, while her strategy benefits all females in the group. Gaining fitness is a relative process, as an individual can only gain fitness benefits at the expense of others of the same species. We must assume that somehow taking the risk to pacify angry males will bring the female more benefits than standing on the sidelines would bring the other females. If the one who takes the risk gains no more fitness from her action than does one who keeps her distance from angry males, the trait to pacify angry males would not be selected.

When a hockey player passes the puck to a teammate to increase the chance of the team's win, while simultaneously lowering the chance of being the star of the game, he does so as a result of a rational strategy. He wants the team to win, which may benefit him more in the long run. Simple natural selection acting on individuals cannot lead to such behavior because natural selection is a short-term response to a current condition.

For now, we will assume that the female chimpanzee who calms down angry males will be remembered by those males, thus gaining more in the long run than females who choose not to pacify angry males. This would be analogous to our remembering the "stars" of athletic competitions. But the larger question—*How the trait of wanting to serve the common good at one's own expense evolved*—remains a mystery. Do we use our rational minds to calculate the optimum strategy between serving ourselves and serving others? Or do our evolved emotional brains lead us to adopt one strategy or another?

Freud claimed that most of us make decisions and choose strategies that we believe will ultimately benefit us. In the course of my (Paddy's) working as a lifeguard and pulling many frightened people from swimming pools and oceans, I faced a choice a few years ago that still gives me pause. My husband, Tim, wanted to go for a swim in a very rough ocean, where red flags had recently been planted in the sand to warn people that entering the water could be very dangerous. We both tend to be daredevils, but I held back this time and told him I thought it would be foolish, and possibly suicidal, to battle those waves. He responded by diving into the next one. I was furious, but also terrified for him. Realizing that I was a stronger swimmer than he, I followed. Within minutes, we were both exhausted and tried to swim in, but the powerful undertow repeatedly washed us back out into the vicious sea before we could anchor ourselves in the sand beyond the breaking point of the waves.

For the first time in my life, I thought the ocean was going to end my days on the planet. I had reached the limit of my endurance and my breath. In desperation, I gave up struggling, totally relaxed and allowed the next huge wave to roll and toss me onto the beach, hoping my lungs would not collapse, nor my neck break. When I realized that I was safely deposited on the beach, I looked for Tim and saw that he was still struggling. I believed that if I dared plunge back into the fury, I would not survive. It was taking every ounce of energy in my body just to breathe. In that intensely emotional moment, I had to choose between staying safe, hoping my husband would comprehend what to do to get back on the beach, or probably dying in an attempt to save him.

My rational mind clicked "on." I thought about my children and grandchildren and knew I could not risk another battle with that ocean. It would be an unimaginable hell to watch the ocean swallow my husband, but it was even more unimaginable for the children and grandchildren to deal with losing both of us. My rational mind also knew that Tim might yet be able to save himself. He did. Only when I was sure we were both safe from the clutches of the waves, my anger over his diving in was unleashed. He probably considered diving in again to escape my wrath.

In my mind, I've relived that moment of choice dozens of times and wondered if I would have gone back into the ocean for one of my children or grandchildren. I'm fairly certain that I would have. What I learned is that our emotional brains and our rational brains can share a tightrope when faced with what we perceive to be life or death choices, but the evolutionary principal of a mother choosing to preserve her posterity usually wins out. It is critical to be as aware as possible of the power love and loyalty wield in our lives—and to know our limits. When faced with circumstances that force us to our personal limits, it is easy to behave irrationally, but it pays to consider the most likely consequences of our behavior and how others might be affected. In analyzing this experience, I've also had to accept how natural selection may unconsciously influence our decisions. In the moment of choice, I was not consciously aware that my children and grandchildren carried my genes, whereas my husband did not.

Our Second Dilemma: Limits on Violence

An interesting theory on the origin of soccer is that the game began during that dark period of British history, between the departure of the Romans and the arrival of the Normans. For much of this time, the British were suffering from repeated raids and invasions by bands of seafaring Vikings. There was a great deal of murder, rape and pillage, as well as genocide (ethnic cleansing) over large parts of the country. However, there almost certainly were periods of relative peace, allowing the population to partake in more relaxing forms of competition. One pastime consisted of two neighboring villages playing a game of "Dane's head."

This competition involved able-bodied men of each village trying to kick a decapitated Viking head into the center of the other village. The starting line was halfway between the villages, and the game could last for days if the two teams were evenly matched. While little is known about the rules of this game (except for the gruesome "ball") it must have been, in principle, quite similar to a contemporary English soccer game. Violence must have been somewhat controlled, passions vented through regulated avenues, and heroes recognized in each village. What interests us most about this version of the origin of a competitive team sport is that it is an intermediate step between serious, lethal warfare and the highly regulated sports events of our modern civilization. Perhaps a move to limit violence began via within-troop social interactions among our ape ancestors, which gradually extended the process of limiting violence through regulated, controlled competitive events among sub-groups of societies.

My family and I (Dolf) lived for a while in Papua New Guinea, which provided me with an opportunity to observe the very strong tribal nature of the New Guinea population. The people openly expressed fear and disdain for members of other tribes. Even villagers who were members of one's own extended tribe, but lived in a valley over the next mountain ridge, were generally mistrusted. The semi-regulated battles between villages were very similar to how I imagine games of Dane's head must have played out. These skirmishes featured body-painted, intricately decorated men running wildly about, shouting and shooting relatively ineffective arrows into one another's general direction. In the end, the combatants expressed a sense of having accomplished something worthwhile and enjoyable. Fatalities were rare, and the skirmishes appeared to have little material effect on the social and economic life of the villages. Even when injuries occurred, the elevated village morale that resulted from having shown the enemy their strength and determination more than compensated for a few arrow wounds.

While in New Guinea, I befriended a local Wampit Valley villager called Yengbong, who had participated in several such skirmishes and deeply resented that he was given a one-year prison sentence for killing a man from their neighboring village during a skirmish. Although his fellow villagers revered Yengbong as a hero, he told me that the village leaders did not want him to kill again. I suspected that they saw him as an overly aggressive troublemaker. The general village population, on the other hand, was perfectly happy with the ongoing payback vendetta. In the past, tribal warfare in New Guinea pitched *ad hoc* armies from neighboring tribes against one another in serious combat, which, in the short run, resulted in many injuries and fatalities. In the long run, it sometimes caused the demise of entire tribes, a kind of genocide.

From our chimpanzee-like ancestors to the present, individuals have experienced a conflict between the drive to aggressively protect their status and the pressure from friends, neighbors or society to tow the line of more peaceful, cooperative behavior. The gradual development of the concept of what we now call the common good, or less euphemistically, the demands of society, is an extremely important aspect of the interaction between our biological evolution and our socio-cultural development. As we stated in the previous chapter, a big part of this interaction is societal control over individual violence. This includes the development of ethical theories, moral codes, laws, law enforcement and social pressure, which have acted for generations as a society-driven version of natural selection. These forms of social control have resulted in a general trend towards both reducing violence and the institutionalizing of violence within structured societies. Nonetheless, differences among societies in this respect remain considerable.

In the first half of the twentieth century, colonial administrator and social anthropologist Bazett Lewis, in *The Murle: Red Chiefs and Black Commoners*, describes how the fiercely warlike Murle people of southern Sudan managed to limit violence within their society. When raiding their neighbors, the Murle used complex strategies and fought with steel spears and knives, but the use of a steel weapon on another Murle during a dispute was considered the basest of crimes. The warrior who returned from battle having killed members of neighboring tribes returned as a hero, but had he injured or killed a fellow Murle with a steel weapon, he would have lost all status in the tribe. If a man had a dispute with one of his own tribe, he fought his adversary with sticks of prescribed dimensions, using leather shields and specified techniques. Draped in special cloaks and lion mane neckpieces, these body-painted men fought their non-lethal duels in public. As described by Lewis, the men would stand…

> … at the ready, with one foot advanced, their arms above their heads and the stick held vertically behind them. From this position they are exceedingly quick at parrying a blow from an opponent by a slight movement of the wrist, which serves to push their own stick forwards and to the left or right side, as the case may be. Downward blows are received on the shield or warded off with the stick; the great art is to be able to stop an adversary's blow and then slip in a quick jab before he has time to recover. ([2], p. 93)

Living with Social Complexities

The Yanomamö of the Amazonian rainforest are a widely distributed tribe of shifting horticulturalists, who gather and hunt in the forest. They live in villages of about a hundred individuals. Until well into the 20th century, they were barely affected by the encroaching actions of national governments and intrusions of western settlers, hunters and other influences. Some villages are more or less egalitarian, with a respected headman loosely in charge, while aggressive, dominant leaders run others more hierarchically. The population of each village is subdivided into patrilineal clans that compete with one another for status within their village. Ultimately, the size of a Yanomamö village is limited by the ecology of the rainforest, which has limited productivity with sparse resources. When ecological constraints start to hurt a village, the people either move lock, stock and barrel to a new site, or split into two groups. The proximate event that leads to a split is usually the eruption of a blood feud between two clans bickering over resources. Because each village is constantly under threat of attack from neighboring villages, making solidarity and cooperation more essential for survival than splitting into factions, these feuds are rare events.

The world of the Yanomamö is one of nearly constant warfare. Small raiding parties, large-scale village-against-village battles and even serious war between alliances of villages are frequent events that lead to many deaths. Women are likely to be captured and brought home as extra wives. In fact, the proximal motive for all this lethal aggression among males is the acquisition of women, who are seen as a limiting resource. For the Yanomamö population as a whole, however, the ultimate limiting resource is the productivity of the rainforest. All this fighting and killing does not create any extra women, nor increase the productivity of the rainforest.[3]

We see two general phenomena interacting in this tribe. First, the stability of the village is of paramount importance for its survival, resulting in a number of generally accepted practices. One is that male-male aggression within a village must be externalized into inter-village warfare. Keeping the lid on internal aggression is achieved through the mediation of the headman (often with the help of his clansmen), and through regulated symbolic, non-lethal fighting between men of different clans over conflict situations. This prevents the splitting of the village, which would result in two weak villages instead of one strong one. Young males are then encouraged to vent their aggression and hopefully satisfy their desire for more wives by raiding neighboring villages.

The second phenomenon expressed by this culture is the control of population density via constant warfare and killing. This assures there will be enough resources for the survivors, although population control is not a consciously acknowledged objective. The alternative would be malnutrition, disease and starvation in a degraded environment. One strange, and to us westerners, offensive, practice of the Yanomamö is a very high level of infanticide, mostly of female infants, which can probably be explained as serving two separate functions. Primarily, it keeps the village population from exceeding its optimal size and the society favors boys, who will grow up into warriors who will defend the village and hopefully raid neighboring villages for wives.

The constant raiding and capturing of women and the battle casualties among men are both the cause and the result of a highly polygynous society. The headmen, even the gentle ones with little political power, tend to have a large number of wives. One headman had 43 children by eleven wives. If nothing else, this indicates to us that the egalitarianism of the Yanomamö has its limits.

The specific social and political complexity of this tribe has for thousands of generations been a self-perpetuating system of rational control and natural selection. Were our ancestors during the millions of years as hunter-gatherers like the Yanomamö? Were our moral and ethical feelings towards our fellow men shaped by similar socio-ecological circumstances?

Kith and Kin

A major complexity of social life is that individuals come as members of a family and family members tend to have a higher sense of loyalty to one another than to unrelated members of their society. In Chapter One, we briefly discussed kin selection, and we will now revisit this important concept in more detail to understand its consequences in the evolution of structured human societies. Why is it that closely related individuals tend to offer more help and support to one another than to unrelated close friends? Relative to the degree of contact and potential conflict among them, lethal violence among kin is rare. Cooperation and trust tend to be more likely among kinfolk than among even the best of neighbors. Can we explain why we generally prefer kin over kith?

Imagine a population of some nonsocial ancestor. If one individual happens to have a rare gene, which makes her want to share a favorite meal with a non-relative, chances are that the other will merely take the offered food and walk away with not so much as a thank you. If he has no sharing gene, why would he reciprocate? The generous individual loses out and natural selection will eliminate the sharing gene. If, however, the sharing gene is more discriminatory and offers treats to close relatives only, the relatives may very well have the same gene, and everyone gains from the sharing practice. In this case, natural selection may very well choose this gene to spread throughout the population. Recognizing kin and extending favored status to them has evolved in several types of animals as an aspect of their social behavior.

Maternal care—and to a lesser degree, paternal care—has evolved independently in the many animals, from insects to mammals. For obvious reasons, such care is usually expressed only towards an animal's own offspring. Only in highly social organisms, and then only at a somewhat lower level, is parental care extended to unrelated juveniles. During my research in the Hudson Bay lowlands near Churchill, Manitoba, I (Dolf) worked in close cooperation with a team of biologists who studied snow geese. These large white birds nest early in spring, and as soon as the goslings are hatched, the parents and their little ones spread out over the salt marshes to graze on the new green vegetation. In a colony of several thousand birds, it is not infrequent to encounter lost goslings, separated from their parents. These little down-covered animals become easy prey to predators, such as herring gulls.

I once watched one of these lost goslings being attacked by a gull. It desperately ran from one goose family to another, trying to get protection. But each time it came close to a parent goose, which was protecting its own offspring, the goose pecked it and chased it away. The pathetic little creature was bleeding from its head where the gull had attacked it, stumbling from what looked like one safe haven to another, only to be cast off again. Soon it gave up, sat down and was killed and ripped apart by the gull. The other goose families paid no attention. The selective allocation of care, such as feeding and protection, to one's own offspring depends on one's ability to recognize offspring as one's own.

It is interesting that some species, such as the honeybee, have evolved a high level of ability to recognize various levels of relatedness, while other species merely assume that a juvenile in one's immediate domain must be one's own. Snow geese do not recognize their eggs as their own except by the simple fact that they are in their nest, but they do learn to recognize their own goslings early after hatching, and will continue to do so even after fairly lengthy separations.

We introduce kin recognition to point out the likelihood that general preferential treatment of kin is an evolutionary elaboration of parental care. Somewhat surprisingly, we humans have a poor innate sense of kin recognition. Nearly all of our recognition of relatives is learned from being told and through interactions with them. This poor sense of recognition creates social problems, especially when a man may doubt his paternity of his wife's children.

Inclusive Fitness

Since we have defined the fitness of an individual as its relative contribution to the gene pool of future generations of the species, it has become standard practice among evolutionary biologists to think of an individual's fitness in terms of inclusive fitness. This means that one's fitness in comparison with other individuals depends on more than one's actual and potential contribution to the next generation, but also, to some degree, on the personal fitness of all relatives with whom one shares many genetically determined traits.

One of the best examples of kin selection in the animal world and its influence on the life strategy of a species is found in several kinds of birds. For instance, in some crows and jays, there can be three apparently adults taking care of a brood of young ones. DNA analysis of some of these threesomes has revealed that one of the adults is a more mature progeny of the other two. The evolution of this behavior of mature, non-breeding offspring helping its parents with the raising of young siblings can be explained by the concept of inclusive fitness. We may wonder why this helper sibling does not find a mate and produce its own offspring. The answer to his question lies in the overall ecology of the species, which demands the ability to attain and defend a territory and the skills to rear a brood under poor conditions. Only the older, more experienced birds have these skills. Helping one's parents not only adds a fitness increment to one's inclusive fitness, but also provides valuable teaching and learning that will pay off in future years.

A human example is my (Dolf's) sister's fitness measured only by offspring would be zero since she has no children and is beyond reproductive age. However, her inclusive fitness is partly determined by my two sons, my brother's four children and his ten grandchildren. Hence, she can increase her fitness by being a good aunt, as indeed she is. Most people prefer to leave their estate to blood relatives, even unpopular ones, rather than to good friends. But there is another side to this coin. In view of what we now know about inclusive fitness, how can we make sense out of within-the-family murders by thinking in terms of inclusive fitness?

Infanticide in the Animal World

During my summers on Banks Island, I (Dolf) had the habit of writing-up the day's work reports and my personal diary entries late in the evening, after the others had gone to their own tents. I enjoyed sitting alone in the communal tent, with a cup of chamomile tea, the oil stove keeping me warm, and the low sunlight streaming through one of the little north-facing windows. One year, I had interesting company for these evenings. About half a mile away, just visible through the window, sat a female snowy owl on her eggs. From her nest, which was positioned on a distinct promontory next to a steep ravine, the owl had an uninterrupted view all around. Her mate was away hunting much of the time, but occasionally, he sat on the ground near her.

Having the owls' nest so near to our campsite was the highlight of that summer. Every second or third day, we visited the nest to check on the owl family. When we approached, mother owl would fly a little distance across the ravine and sit on a rock watching us. She seemed unperturbed at our incursion into her nest site and returned to her brood immediately after we walked away. She laid seven eggs, but unlike most bird species, which only start to incubate after the last egg is laid, she started incubating as soon as she laid her first egg. This had an interesting result: the adorable fluffy pale-gray chicks hatched in sequence over a period of about twelve days, so that the oldest chick was already quite large when the last one hatched. Unfortunately however, we did not witness the development of a happy, peaceful owl family. Only the oldest two

110

chicks fledged; the other five were killed and eaten by their older siblings. Fratricide of this type is fairly common among predatory birds. It is an integral part of the life strategy of the species. In a year of abundant lemmings, a female snowy owl can easily get enough to eat to produce and lay an egg every day or two. However, an abundance of lemmings in early spring does not necessarily translate into an abundance of lemmings later in the summer. In years with season-long abundance, six or seven chicks may get to eat so much that the urge to kill and devour a younger sibling is unlikely to surface. In years of declining abundance of prey, a nest with similar sized chicks would probably fail, as there would not be sufficient food, and the chicks would fight and injure one another.

In my Arctic and sub-Arctic research sites, bears were a frequent intrusion into our daily routine. Like all predatory omnivores, when a bear spots another animal in its proximity, it has to balance the food value against the risk of trying to kill the potential meal. Most bears (but not all!) consider humans very risky prey, especially when we are in a group. But even prudent bears can become dangerous when very hungry, and sows with cubs can be dangerous when they sense that we are threatening their cubs. Male bears, being generalist predators, will try to kill and eat any animal that they consider relatively safe to tackle, including cubs of their own species, if they encounter the cub far enough from the sow. Since male bears roam over wide areas and form no pair bonds with females, the chance that the cub will be their own offspring is so low that on the balance sheet, killing cubs for food is a fitness-enhancing strategy. A good meal gives a larger fitness advantage to the bear than the average fitness price paid by the small risk of actually killing his own offspring and/or by getting injured by the cub's mother.

Infanticide is also practiced among mountain gorillas. But their story is much more complicated because we are dealing with highly social animals. Dian Fossey, who has studied mountain gorillas for several years in Rwanda, described how relatively young silverback males, who are in the stage of trying to establish their own troop of breeding females, use a strange but effective strategy. When such a male attacks an established family troop, he focuses much of his aggression on the old silverback, but he also tries to kill infants from the established troop. If he succeeds, chances are that after the confrontation, the mother ape who lost her infant will abandon her troop and join the young silverback. This is obviously advantageous to the young male, but what is in it for the female? The status rank order of females in a gorilla troop is determined by the seniority of the females; those who have been in the troop the longest have the highest status. If a young silverback kills the infant of a low status female in the troop he attacks, the female gains status by joining him as one of his earliest females. This strategy does not work in chimpanzees, because they form larger, multi-male troops, with a very different female rank order system. Since the females gain nothing from switching troops, males from other troops do not kill infants for the purpose of gaining reproductive females. There is, however, evidence of males killing infants as a by-product of group fighting, or simply for food.[4]

Infanticide in Primitive Societies

Most of us can understand the role of infanticide in animal species, but in our supremely social species, we would expect to find a strong aversion to harming children—especially our own. Unfortunately, many primitive human societies practice infanticide frequently as part of their social and individual life strategies. In many cases, it is the only safe solution to unwanted pregnancies and births. Infanticide as a conscious strategy has been well described by anthropologists, Paul Bugos and Lorraine McCarthy, for the Ayoreo, horticultural and seasonally nomadic foragers of the Bolivia Paraguay border region. Until the mid-twentieth century, the Ayoreo had firm birthing traditions that allowed mothers, with their attending women, to make an immediate decision as to whether to keep the baby or not. If the economic conditions and/or the quality of the baby were such that the decision was to abandon the baby, it was quickly buried alive, without having been touched by human hands. Bugos and McCarthy stress that infanticide, however prevalent, is considered a regrettable tragedy that mothers of the euthanized babies do not want to talk about. Women who early in their lives have abandoned some of their infants are frequently described as proud, loving and caring mothers for the children they eventually raise.[5]

During my stay in Papua New Guinea, I (Dolf) conducted a research program on rainforest biodiversity in the lower Purari River valley. I had the opportunity to come to know some of the local river fishing/gathering people. These people do not have villages or houses; by day, individual families live in their dugout canoes, and at night, they camp along the shore of the river. Each couple had either one or two children with them, and when there were two, they were at least six years apart, which caused me to wonder what kind of birth control they practiced.

I remember well the close, loving family relationships that I observed—especially when one such family spent a night camping on a clear bit of shoreline near my camp. The father was telling his two children a bedtime story, as the sun was setting beyond the forest on the opposite shore of the untamed Purari. Later, on my return to the fringes of civilization, I mentioned my experience to an Australian merchant in the coastal village of Baimuru. He scoffed at my naivety and replied, "Those people are infanticidal savages."

In primitive societies, which function in small units and have rapid turnover economies without reserves, unwanted children have nowhere else to go, making infanticide a remarkably common solution. In contemporary societies with well- developed social systems based on ethical principles, unwanted pregnancies are frequently terminated early and safely, while unwanted babies are often brought up by relatives, or adopted. Infanticide may seem a cold-hearted perspective and strike the reader as shockingly inappropriate for the human species, which, in a social/ethical sense, it is. If we want to understand why infanticide is as common as it is, we must take a hard-nosed, scientific look at the problem.

By and large, infanticide is practiced for two overarching ultimate reasons: the mother's long-term fitness strategy and the mother's pair-bonded male partner's reproductive fitness. These two reasons often interact in unexpected, complex ways, as in the mountain gorilla. In *Homicide*, Daly and Wilson argue convincingly that the incidence of infanticide can best be explained by the personal and socio-ecological conditions of the mother at the time of the child's birth and how these conditions affect the balance between cost and value of the child to the mother's fitness. For societies with limited resources and few social safety nets, one of the most important criteria leading to infanticide is the fitness quality of the infant. Misshapen or obviously unhealthy infants are often killed immediately after birth, as their chance of survival to healthy, productive adulthood is extremely low. From an economic standpoint, keeping and caring for such an infant would probably be a waste of effort, time and resources. The mother also has to consider the alternative investments to which she is already committed. For instance, are there older healthy siblings who need the parents' time and resources? Or, is the mother young enough to be able to count on future reproduction? Does the mother have the support of a husband or other family members she can count on? If not, especially if the mother is very young, abandoning the child or even killing it may be her best strategy. For the husband, it is of utmost importance to believe that the child is his. A child is much more likely to become a victim of infanticide when the husband suspects, or knows, that he is not the father. In several primitive societies, it is common practice to kill a widow's children before she can acquire a new husband.[6]

This gruesome system of family planning practically disappeared with the coming of agriculture, when children became assets for the farmer. The value of their labor exceeded the cost of their maintenance. Infanticide was the last within-the-group lethal violence that was externalized beyond the pale. The flip side of the coin was that rapidly growing agricultural populations, inevitably exceeding their resources, began to wage war and genocide—which still happens with alarming frequency.

Hunter-gatherer and low-level horticultural societies have existed for many millennia as stable, sustainable entities. Such stability is often romantically ascribed to some kind of innate understanding by primitive people of how to live in harmony with nature. The reality is much less romantic. Most primitive peoples survived as social groups either because they lacked the technology to over-harvest their resources (resulting in frequent periods of starvation, but rarely complete collapse) or because they curbed their own population as a byproduct of various forms of homicide. We have already seen how the Yanomamö survived via both inter-village warfare and intra-village infanticide. The Ayoreo, the Purari fishermen and other primitive societies solved their ecological problems by practicing infanticide within the family.

Infanticide in Civilized Societies

For the most part, contemporary human societies are comprised of people who have lived for many generations under highly structured, culturally directed social conditions. This reality has imposed on us rigorous selective pressures long enough for us to shape ourselves into a different kind of being. Far more than our hunter-gatherer forebears, we are willing to adjust our individual behavior according to socially imposed rules. Added to this evolved tractability is a cultural ethic of obeying laws that we are taught from childhood, which further influences us to bow to the authority of society and its leaders. The result is that we have accepted the externalization of lethal violence as the right and normal way of organizing our world. In other words, mankind has been caught in an evolutionary pull to kill and to wage war. This ethic is so pervasive that it comes as a surprise to most people when we explain homicide within our society in terms of evolutionary fitness rather than in terms of psychopathology or crime.

In a study of over a thousand cases of child murder in Canada from 1960 to 1980, Daly and Wilson have made a number of predictions based on evolutionary theory and our understanding of reported infanticide in the anthropological literature. Theory states that a child's fitness value to parents goes up with its age, because the older a child becomes, the less chance remains that he or she will die before producing children. Also, the expanded cost of bringing the child to maturity becomes progressively less as the child gets older—not considering the cost of a university education. Hence, Daly and Wilson predicted that a child's chance of being murdered by his or her parents would decrease with age, but that this would not be the case when the murderer is unrelated to the child.

The analysis of the data strongly supports their theory. If, in our evolutionary history, young unmarried mothers were driven by economic desperation to kill their infants, we would expect a measurable trace of that inclination to persist in the innate component of contemporary human behavior. Daly and Wilson found that a disproportionate number of infanticides were committed by young, unmarried mothers. Since stepparents are not genetically related to their stepchildren, these children have no fitness value for the stepparents. That simple fact suggested to Daly and Wilson that children would be at greater risk of abuse, neglect and murder in families with a stepparent than in families with only natural parents. The statistical facts supported their prediction; the fairytales about vicious stepparents are not entirely based on irrational fears.[7]

Am I My Brother's Keeper?

When my (Dolf's) two children were of primary school age, they constantly quibbled over who got the biggest pancake, who had the best toy, who was allowed to sit in the front seat of the car, etc. It was a classical case of sibling rivalry, which demanded frequent intervention by their parents. It was therefore a surprise to us to discover that on the schoolyard, the older one was the loyal protector of his younger brother. No bully could taunt or lay a hand on the little one without having to deal with the wrath of his big brother. This should not have been a surprise to us, since we knew that brothers share fifty percent of their genes. Therefore, a large part of their inclusive fitness was invested in the other, and therefore, was worth protecting. But, at home, where we parents did the protecting, the boys, being in direct competition for the nepotistic component of their fitness, responded naturally to the other fifty percent of their genes. It is a small-scale example of the hockey player's dilemma of choosing between competition and cooperation? What makes the situation even more complex and interesting is that within a family, there is also parent/offspring rivalry and father/mother rivalry, and all three rivalries interact cooperatively and exploitatively.

Under most circumstances, families survive as units simply because kin selection has favored cooperative behavior over competition. But, in some circumstances, selection has left an opening for serious competitive behavior, including violence within the family. The family is the smallest unit of social activity, and, as with all social units, its borders are vague and stretchable. To a large extent, its cohesion is based on kin selection, but whereas parents and children are closely related, sharing a high level of inclusive fitness, the parents are usually unrelated to one another. They are bonded through very strong common interests. The extended family consists of many individuals with variable levels of relatedness. At the limits of what is considered family, relatedness overlaps gradually with the concept of clanship.

116

Through intermarriage, many families form alliances, while other families may end up feuding. As we have seen, infanticide in socio-economically primitive societies is mostly a matter of parental concern, whereas, in more advanced societies, it becomes a societal concern. Other forms of violence within the family, such as fratricide and parenticide, are more complex in their social context, as they are an inclusive fitness concern. Unlike young children, older individuals always have an inclusive fitness value to all of their blood relatives. When a non-relative kills a family member, it is seen as a matter of concern for the entire family. When a close relative kills a person, the situation is much messier, as some relatives will lose inclusive fitness, while others may gain it.

In Chapter Three, we met Nathan and Donna and their two children. We learned more about Donna's family and can assume that Nathan also had an extended family. Donna's murder affected both families. The immediate psychological, emotional and behavioral responses of the family members are an evolved response to the effect of the murder on everyone's inclusive fitness. Donna's parents lost not only a daughter, but also any hope of additional grandchildren by their daughter. Nathan's arrest and incarceration also robbed his parents of any further grandchildren. We discussed why Nathan would commit such a murder, but we must revisit this situation briefly. Spousal murder is the most frequent homicide within the family for a simple reason: husband and wife are not related. What keeps them together is their common interest in the children. If that common interest is absent or weak, and other fitness investments exist or are expected, the bond between a couple may become a liability to one or the other. The chances of severing the bond through violence are often increased. In Nathan's case, he responded in an innate, evolved manner to his perception that Donna was a barrier to his potential for future reproduction. With her murder, he reduced the inclusive fitness of Donna's family members, including his own children's, which illustrates the selfish nature of evolved behaviors. The fact that his incarceration reduced the inclusive fitness of his family members was probably not part of his subconscious cost/benefit calculation, because our legal system is an expression of the fish-tank phenomenon, discussed in Chapter One.

The case of Deborah's murdering her husband is also a spousal murder, but more complicated because it involves the nepotistic component of fitness. We already know that Deborah had children by a previous husband, so we must assume that there are three extended families involved, each of which has an inclusive fitness stake in this affair. Had Deborah not been caught, she, her children and the extended families of her and her previous husband could have gained wealth and status (fitness) while robbing the murdered man's family of its rightful share. In both of these cases, Nathan's and Deborah's, we should notice an interesting resolution: the homicidal spouse is caught and convicted, but the aggrieved families are not compensated for their loss. This is a fairly recent practice in human society. In practically all more primitive societies, post-homicide resolutions are based on a tit-for-tat concept or on economic compensation. In these societies, families of someone who has been murdered, or even accidentally killed, demand the death of a member of the killer's family, or payment, which increases their nepotistic component of fitness.

The same is often true in the animal kingdom. With the snowy owls, it was the parents' strategy that allowed the older chicks to kill their younger siblings at times of food shortage. The situation is different with the bald eagle. A study in Saskatchewan showed that eagles usually lay two eggs, three days apart, which means that they hatch three days apart. The two sibling chicks fight in the nest for parental attention in the form of feeding, cleaning and brooding. This is understandable sibling rivalry, but the fighting can be intense enough that it often leads to the death of one chick. This is not a parental strategy, but fratricide. In fact, the eagle parents, like most human parents, put effort into keeping the lid on the offspring's fighting. The older chick has a distinct advantage, especially if it is a male because the males grow faster. The battles are more evenly matched if the firstborn is a female. Researchers have discovered that the eagles lay significantly more female-first than male-first clutches, which is sure proof that they want to raise both chicks. It remains a scientific mystery how these birds are able to manipulate the sex allocation of their eggs.

Human overall fitness is not just a matter of genetic relatedness, but also includes an important nepotistic component. In hunter-gatherer societies, this is of little importance, except for individuals, such as

118

sons of headmen, who could inherit status from their fathers. But heritable lineages are rare in these societies. With the coming of agriculture, land ownership and associated wealth created a new competitive situation for family members. Murdering one's father or brother in order to inherit the family land and wealth could become a fitness strategy, often used as a plot in literature and theater. Daly and Wilson's exhaustive study reveals a long history of internal family violence. In primitive societies, this was of little concern to the group and the problem remained within the family, but intra- family feuds became another matter. It was important for communities to have specifically prescribed methods for suppressing family feuds since they had a negative effect on the common good. This inevitably led to the externalization of violence. Land is limited in an agricultural community, so brothers or fathers and sons face an ever-diminishing land per family ratios. They often fight to the death in order to claim the family farm.

In modern, structured state societies, the family has lost its power to deal with family violence, since the state enforces the laws and punishes the transgressors. With this ever- increasing trend of a society's governing body making the decisions about how to handle violence, legal lethal violence can be moved onto the battlefield. But why would men and women be willing to go to war and take the risk to have to kill complete strangers or be killed by them? To begin to answer this question, we must return to the evolution of our concept of altruism.

The Two Faces of Altruism

What is the origin of altruism? Social animals frequently appear to make minor sacrifices for the benefit of fellow troop, pack, herd, or pod members. Monkeys groom one another, wolves share food and muskoxen wait up for a female delayed by giving birth. All this may appear to be altruistically motivated, but it is better understood as cooperative behavior. If one group member does not groom, share or wait-up, the minor effort and time saved would not outweigh the cost of being disliked and eventually shunned by the others. Even the female chimpanzees who risk pacifying angry males will probably be rewarded by other group members in social currency, such as trust and protection.

This is a more advanced form of cooperative behavior, since it involves a major risk and a longer time span between taking the risk and getting the reward. It works because the reward is diffuse and does not have to come from all members of the group. Selfish individuals can survive in the group without taking risks and without showing respect for the risk-taker, but they will not receive others' respect, or other forms of social reinforcement. We can understand this kind of risk-taking on the part of female chimpanzees as, on the average, a naturally selected behavior.

Social rules and their enforcement are the environmentally imposed driving force between social and natural selection. It is possible, but questionable, that the beginnings of such rule making and enforcement can already be found in chimpanzee society. But, in the human species, it is a very strong evolutionary force. We do not believe that true altruism has evolved in our ancestors because of group selection, which is the preferential survival of groups that have some altruistic individuals over those groups which do not. Nor do we believe that reciprocal behavior is a reasonable explanation for the evolution of altruism. The theory that our ancestors learned who was a reliable reciprocator and who was not leaves out the strategy of appearing to be reliable—until it becomes a matter of life and death. War reports are full of cases where the soldier who had saved an injured comrade under heavy fire was, at a later date, abandoned by the very one he had saved. The only convincing explanation for the evolution of true, self-sacrificing altruism is harsh socio-cultural enforcement. The shirker and the cheater are ostracized or penalized; altruists and their close kin are rewarded.

Many countries have "Good Samaritan" laws, which can result in severe penalties for those who do not help accident victims, while many countries award medals-of-honor for heroism to those who do help others when they themselves are at risk. Rewards for kin of those who perished while helping others contributed considerably to strengthen this trait, because it made self-sacrifice for the common good a strategy that would enhance one's inclusive fitness.

We usually think of altruism as an admirable trait in individuals who take severe risks to help others in need. In most cases, the young soldier who goes off to war to defend his society against the evil enemy is an instant hero and usually goes to war feeling proud of his decision.

Once on the battlefield, other powerful emotions enter the soldier's mind. Some of the most risky, heroic actions are driven more by hate and anger than by a sense of duty. There are many stories of men who, when seeing their buddy killed, throw all caution to the wind and single-handedly charge a well-armed enemy position. Love for their comrade becomes stronger than self-love, and in an intense moment of passionate love and passionate anger, rationality loses.

Complicating the picture still further is the reality that the same young men, who readily go to the battlefield at times of war, are also the most apt to engage in highly competitive activities when building a career and pursuing other life goals. What makes the situation especially difficult is that social pressure on individuals to serve the common good places constant, but opposing, pressures on competitiveness to achieve one's goals in life, which can undermine the common good. Balancing cooperation and competition in our lives is a constant challenge.

Misguided Altruism

When people become frightened, we are vulnerable to a darker aspect of altruism—a willingness to unquestioningly follow a powerful leader who promises to protect the common good. We have seen again and again, that leaders do not have to be despots to dominate and manipulate the people they lead. Many current democracies are not led by tyrannical regimes, but by politicians who manage very effectively to manipulate the populace. Winston Churchill realized early in the Second World War that the British people were not fully convinced war was necessary or unavoidable. But, via organizing a major propaganda campaign, which effectively utilized the aerial assault on Britain by the Luftwaffe, Churchill soon had the people solidly behind him. Unfortunately, this phenomenon issues from the dark side of altruism. It is an evolved part of our nature to save the group by blindly following a strong leader—even accepting the need for sacrifice of life and of fundamental ethics. History is filled with unnecessary, propaganda-initiated wars—from Athens' invasion of Sicily to America's invasion of Iraq—fought by people who offered themselves to do the bidding of leaders who allowed their greed and fear to incite a call to arms. Suicide bombers, driven by the passion of religious and political indoctrination, exhibit an especially sinister form of misguided altruism. These phenomena warn us that altruism can result in destructive thinking, which initiates evil, just as easily as it can initiate good. To prevent this, we must be willing to develop and use our higher consciousness, which can help us survive with more compassion and less violence.

References:

Q. Paine, Thomas. In Baker, Daniel B. (Editor) *Power Quotes*. USA: Visible Ink Press for Barnes and Noble Books, 1992.

1. de Waal, Frans. *Good Natured: The Origins of Right and Wrong in Humans and Other Animals.* . Cambridge, Massachusetts: Harvard University Press, 1996.

2. Lewis, Bazett. *The Murle Red Chiefs and Black Commoners.* Oxford, UK: Clarendon Press, 1972.

3. Chagnon, Napoleon. *Yanomamö: The Fierce People.* Fort Worth, Texas: Harcourt, Brace Inc., 1997.

4. Fossey, Dian. *Infanticide in Mountain Gorillas.* In Hausfater and Hrdy, (Editors) *Infanticide: Comparative and Evolutionary Perspectives.* New York: Aldine de Gruyter Press, 1984.

5. Bugos, Paul E. and McCarthy, Lorraine M. *Ayoreo Infanticide: A Case Study.* In Hausfater, Glenn and Hrdy, Sarah (Editors) *Infanticide: Comparative and Evolutionary Perspectives.* New York: Aldine de Gruyter Press, 1984.

6. Daly, Martin and Wilson, Margo. *Homicide.* New York: Aldine de Gruyter Press, 1988.

7. Ibid.

Call to Arms

Please try to understand this. It's not an easy thing to hear, but please listen.
There is no morality in warfare. You kill children.
You kill women. You kill old men. You don't seek them out,
but they die. That's what happens in war.
~ Paul Tibbets

Paul Tibbets was the "flyboy" who piloted the plane that dropped the atomic bomb on Hiroshima. Within eight minutes, an estimated 140,000 people were killed. The beginning of the end of World War II was set in motion. The western Allies viewed the destruction caused by this bomb as an act that would shorten the agony of our soldiers. Little consideration was given to the agony caused to others, "the beasts," as U.S. President Truman called the Japanese when he tried to justify the use of such a weapon.[1] What happens in the human psyche that allows us to perceive "the enemy" as "the beasts," or to perceive ourselves as better than others?

An Early Awakening

In the summer of 1944, during World War II, I (Dolf) lived with my parents, brother and sister in Nazi-occupied Europe. Our home was in a small town in the Netherlands near the eastern shore of a large lake. The summer was long, warm and deceptively peaceful for my brother, Arvid, and me. At ages thirteen and eleven, we did not consider ourselves vulnerable to the war. We were blissfully unaware of our father's heavy involvement in the resistance movement. On clear sunny days, we would lie on our backs in the grass, passing the binoculars back and forth, as we waited for the daily show to begin.

The first signs were far to the west, beyond the lake. What looked like strange clouds, which would separate into streaks of vapor, rose vertically from the horizon. Before we were able to hear them, we could see the heavily laden American bombers at the tips of these contrails on their way "to teach those Nazis a lesson." We knew the planes: Flying Fortresses and Liberators, accompanied by Thunderbolts and Lightnings. As they flew closer, the sound of many hundreds of heavy engines was deafening. We dreamt of being up there in one of those planes with our American heroes, roaring through the thin atmosphere on our way to smash the German defenses and weapons industry. In my naïve mind, I saw the war as a romantic show of heroes and villains.

One day in late July, after an especially heavy flight, which we later learned had targeted Hamburg, Arvid and I waited for the westward return flight of our heroes. Sooner than expected, we spotted a single Liberator returning ahead of the main formation. Flying significantly lower than normal, its fuselage had a number of gaping holes and the end of its right wing was gone. Just as the big plane was at the nearest point to us, we saw two Messerschmitt fighters diving toward and streaking past the damaged Liberator. Within seconds, we heard the noise of machinegun fire and the screaming of engines pushed to the limit. Our heroes were in combat!

We wanted to see more, but it was over as suddenly as it had started. As the Liberator continued on its course, a streak of black

124

smoke intensified from close behind. We watched the Liberator explode in a fireball and a mass of smoke, out of which chunks of still burning remnants were crashing down to Earth. Only four parachutes drifted down in the blue sky. The battle was over. Our heroes had lost. I remember vividly feeling intense trembling and nausea when I realized that the other six crewmembers had been blown apart and burned. War had shown us its real face. I knew, even then, that pilots and soldiers, Americans and Germans, were young men—young men who loved and were loved. Only slightly older than I, but bound by the rules of war, they were ripping each other apart.

What I observed that summer afternoon drove the cruel truth of war deep into my mind and soul. I had been indoctrinated with the veiled propaganda of the Dutch nation and taught to distrust German propaganda, which we secretly mocked. As I matured towards the end of the war, I began to recognize the hidden lies embedded in the nationalistic fervor of the Dutch people, which tried to make us believe that all Germans were evil and that all Dutch underground workers were heroes. The propaganda had distorted my sense of ethics. As an adult, I came to realize that what Germany began was inexcusable and barbaric, and that what my heroes of 1944 did was also barbaric. The truth is that war makes barbarians of us all.

It's strange how truth emerges eventually, regardless of the means we humans use to try to hide or distort it. When it comes to war (and love, for that matter), we obfuscate facts to delude others as well as ourselves. The saddest aspect of this is that by the time the truth emerges, enormous damage has already been wrought through the distortions and lies. War cannot be undone. The millions who have been killed and maimed cannot be restored to life and health. And peace rarely lasts for long following a war. There have been only twenty-nine years in recorded history during which a major war was not being fought somewhere in our world. So, how do these fatal conflicts begin?

Creating Enemies

For a war to begin there must be someone to blame, someone to fight, someone to conquer. We must create an enemy to feed our dark side of fear and hate. As most adolescent boys are aware, we must call forth, at least in our imagination, heroes and villains. According to author Sam Keen in *Faces of the Enemy*, the process of creating an enemy is as follows:

Start with an empty canvas.

Sketch in broad outline forms of
men, women, and children.

Dip into the unconscious well of your own
disowned darkness
with a wide brush and
stain the strangers with the sinister hue
of the shadow.

Trace onto the face of the enemy the greed,
hatred, carelessness you dare not claim as your own.
Obscure the sweet individuality of each face.

Erase all hints of the myriad loves, hopes,
fears that play through the kaleidoscope of
every finite heart.

Twist the smile until it forms the downward
arc of cruelty.

Strip flesh from bone until only the
abstract skeleton of death remains.
Exaggerate each feature until man is
metamorphosed into beast, vermin, insect.

Fill in the background with malignant
figures from ancient nightmares—devils,
demons, myrmidons of evil.

When your icon of the enemy is complete,
you will be able to kill without guilt,
slaughter without shame.

The thing you destroy will have become
merely an enemy of God,
an impediment to the sacred dialectic of history.[2]

Once this image is created and nurtured through propaganda, we become easily convinced that the image is accurate. Fear feeds our belief that an evil enemy exists, who must be destroyed to ensure our survival. Most of us, on a personal level, cannot imagine that we are capable of participating in such a process of vilification. We prefer to believe that we will be sensible and rational when it comes to recognizing an enemy, and that we would support war only as a last resort. Although we are aware that war legalizes unbelievable cruelty, most of us claim that we would never tolerate war crimes, regardless of who committed them. Unfortunately, the facts tell a different story. Few of us remain rational when an international crisis develops, just as few of us keep a cool head when a personal crisis develops. Whenever we are hurt or fearful of being hurt by someone, we are quick to begin the process of vilification. Once engaged in this process, our ability to be rational is diminished by fear.

Fear lessens our ability to be reasonable and to negotiate. It is also highly contagious. When fear spreads through a population, a level of panic results that seems to paralyze our ability to behave morally and ethically. As hatred and fear escalate, it becomes easier to believe that war is the only "reasonable" answer, even though there may still be time to negotiate a more peaceable solution. Once at war, we become appalled by the cruelty of the enemy, the one we blame for igniting our fear and causing war. We also become blind to the horrors that "our side" may be inflicting on our enemies.

It's time to recognize how morals we claim to hold dear shift—or totally disappear—during war. In 1937, the German Luftwaffe introduced the world to terror bombing by flattening Guernica, a small town in northern Spain, to assist the Fascist side in the Spanish Civil War. Guernica was of no strategic or military importance, but Hitler chose to use this town as a testing ground for his latest strategy—fire from the sky. It was a market day, so the center of town was filled with farmers and shoppers. After three and a half hours of strafing and bombing, the city was completely destroyed. Four days later, the May Day parade in Paris swelled to over a million people protesting the barbarism of the Germans' raid. The civilized world was genuinely aghast, not only at the actual bombing, but even more so at the concept that a government, far away from the war, targeted innocent civilians.

Pablo Picasso, then exiled in Paris, painted Guernica on a huge canvas, which has become the most famous anti-war art of the twentieth century. He did not paint planes and bombs of fire, but a black-and-white image of a wailing mother holding her dead child in her arms, a dead man holding his broken sword, a snorting bull, a screaming horse, another woman on fire, another falling to her knees, and another shining a light on the scene from her window. This painting was exhibited at the World's Fair in Paris later that summer. It was reported that people had a difficult time looking at it, because it made them fear that everything they loved could be easily destroyed by other human beings. Their fear was justified. The painting was prophetic. Two years later, Hitler invaded Poland with the same bombing strategy.

Guernica is a painting by Pablo Picasso, in response to the bombing of Guernica, Basque Country, by German and Italian warplanes at the behest of the Spanish Nationalist forces, on 26 April 1937, during the Spanish Civil War. The Spanish Republican government commissioned Pablo Picasso to create a large mural for the Spanish display at the Exposition Internationale des Arts et Techniques dans la Vie Moderne (1937) Paris International Exposition in the 1937 World's Fair in Paris.

Guernica shows the tragedies of war and the suffering it inflicts upon individuals, particularly innocent civilians. This work has gained a monumental status, becoming a perpetual reminder of the tragedies of war, an anti-war symbol, and an embodiment of peace.

Wikipedia, the Free Encyclopedia

Prior to Hitler's invasion of Poland, the world learned that the Japanese air force had similarly bombed Chinese cities. Once again, many nations expressed abject horror at the inhumanity of the acts. The leaders of the United States and England referred to the bombings as "crimes against humanity." Soon after the onset of World War II, the Luftwaffe began an offensive of terror bombing, targeting first Rotterdam in the Netherlands killing over a thousand people. Then, they bombed the English city of Coventry, in a raid the German government cynically called "Operation Moonlight Sonata." Western democracies were shocked and outraged by this new strategy of aerial warfare.

However, as hatred of the enemies escalated, so did the willingness of the Allies to use any means available to destroy them—civilians included. Early in 1940, Great Britain's Royal Air Force conducted a few bombing raids against German cities, claiming to target only military and industrial installations, even though they realized that the workers in the factories were old men, women and even some children, as all the young men were fighting on the front lines. Since the RAF missed more targets than they hit, hundreds of unintended murders were committed. The German air attacks that had been declared immoral by Winston Churchill were not nearly as destructive as the firestorms he sanctioned. In July and early August 1943, the RAF killed almost as many civilians (estimated around 50,000) in raids against Hamburg as the English would lose during the entire war. Following these massacres, an observer wrote that human torches ran down the streets screaming while tiny children lay like fried eels on the pavement. [3]

On March 9 and 10, 1945, the largest single-day killing in the history of the world up to that date took place when the U. S. Army Air Force flew a force of 334 B-29s, each loaded with over eight thousand pounds of napalm incendiaries, over Tokyo. General Curtis LeMay, who planned the attack, realized the immoral aspects of what was about to happen, and said to his pilots in the briefing prior to take-off:

No matter how you slice it, you are going to kill an awful lot of civilians. Thousands and thousands. But we are at war with Japan. We were attacked at Pearl Harbor by the Japanese. Do you want to kill Japanese or be killed by them? All war is immoral. If you let that bother you, you are not a good soldier.

When the bombs were dropped, thousands of white-hot fires lashed out like dragons' tongues destroying everything they touched. One pilot reported that he felt as though he were looking directly into hell. The final estimated number of deaths stands close to a hundred thousand—mostly civilians. A Japanese student, who managed to survive by plunging into a river and staying submerged for intervals, wrote afterwards that on that night he learned to hate Americans from the bottom of his heart.[4]

But we need more than an enemy to wage a war. War also requires a leader who is willing to issue the "call to arms." This leader, often already corrupted by invested power and blind to his or her desire for more, makes a conscious decision to forego peaceable ways of solving problems.

From Hail Caesar to Heil Hitler

> Beware the leader who bangs the drums of war in order to whip the citizenry into a patriotic fervor, for patriotism is indeed a double-edged sword. It both emboldens the blood, just as it narrows the mind. And when the drums have reached a fever pitch and the blood boils with hate and the mind has closed, the leader will have no need in seizing the rights of the citizenry. Rather the citizenry, infused with fear and blinded by patriotism, will offer up their rights, and gladly so. How do I know? For this is what I have done. And I am Caesar.[5]

There is no record of Julius Caesar's saying these exact words. But, the voice of history tells us that not only Caesar, but Alexander the Great, Napoleon, Hitler, Stalin, George W. Bush and legions of other leaders have manipulated a frightened people into unnecessary wars against a perceived enemy. It is probably not a coincidence that the above quote of unknown origin has been circulating on the Internet since late in 2002, when the President of the United States began to bang the drums of war to convince the citizenry to go to war against Iraq, referring to Iraq as a part of "the axis of evil." The most pernicious kind of manipulation becomes possible when people feel threatened, as the Americans did after the terrorist attacks of September 11, 2001 and as the Germans did in the 1930s. We must remember that Hitler was democratically elected before he took on a more dictatorial role.

When people feel threatened, two traits come to the fore. The first is a desire to be led by an effective, powerful leader, who promises to protect the common good. The second is a willingness to make sacrifices for the perceived safety of one's community or country. If a powerful leader fuels the citizens' fears with suggestions of further threats from a real, or imagined, enemy, those citizens can be easily exploited. Throughout history, leaders and their henchmen have utilized such opportunities to empower themselves and to embroil their nations in unnecessary wars. The process can be likened to a group of frightened sheep being led by a wayward sheepdog into the wilderness.

The Deadly Duo: An Enemy and a Warmonger

In a powerful speech during the spring of 2003, U. S. Senator Robert Byrd explained to the world exactly how the escalation of fear of hatred lured the American people into believing that an unprovoked attack on Iraq was justified to protect the world from terrorism. Addressing the Senate, Byrd explained that following the horrific attacks of Al Qaeda, master-minded by Osama bin Laden, on the World Trade Center on Sept. 11, 2001, America's fantasy that it was invincible was cracked wide open. Americans discovered that we were no longer safe just because we were wealthy and powerful. Fear invaded our minds and hearts, making it effortless for our president to plant the perception that evil empires lay in wait to destroy our good country. In fact, President George W. Bush had other grievances against Iraqi Dictator Saddam Hussein, including an assassination attempt on Bush's father. Saddam also controlled a country with one of the largest oil reserves in the world.

So the Administration immediately set about manipulating the citizenry into pairing Osama bin Laden with Saddam Hussein, who actually had nothing to do with the 2001 attacks. President Bush and his cabinet invoked every terrifying possibility imaginable from germ warfare to the nuclear destruction of all major American cities by weapons of mass destruction, which they warned were ready to be fired in our direction at a moment's notice. They professed that the only way to prevent this disaster was to attack the oil-rich, sovereign country of Iraq. Unfortunately, the U. S. media cooperated fully with our government in spreading lies and fear.

The rest is history. No weapons of mass destruction have ever been found. Thousands more American military, Iraqi military and innocent Iraqi citizens have been killed and maimed than were affected by the attacks of September 11, 2001. It has become painfully clear that Iraq was not an immediate threat to the United States, and that capturing Saddam Hussein has not stopped the murder and devastation of a people, whom we were told we would liberate. Fear and hatred have escalated throughout the Mid-east and the United States to the point that we are presently in danger of another major war. According to Senator Byrd, the cause of freedom and world peace has probably been set back at least two hundred years by this war. Saddest of all, this same manipulative process of warmongering has been happening since the beginning of recorded history.[6]

After the war against Iraq was well underway, I (Paddy) was invited to participate in a two-week tour of Turkey. Prior to traveling, I received several documents from the U.S. Department of Tourism explaining dangers lurking in Turkey, because it is bordered by Iraq and Iran. Though the documents were carefully worded, there once again was that subtle pairing—that Turkey was so close to the "axis of evil" that it could be considered a first cousin and should be feared. Instead, the documents served to remind me that Turkey opposed our attacking Iraq, and that some angry Turks might well take out their fury on visiting Americans.

What actually happened in Turkey was that our group met friendly, welcoming people eager to share their beautiful country with us. Frequently, while walking down the streets of Istanbul or touring the ancient ruins that are being excavated and restored, we encountered school children who would smile and call out, "Welcome to Turkey!" They often followed their greeting with the question, "Are you American?" If we responded in the affirmative, they would ask, "Do you vote for Bush?" and quickly add their negative opinions of Mr. Bush. We visitors realized that these children might be aware of things that our citizenry sees less clearly. Perceptions of truth vary according to cultural heritage, teachings and propaganda. Neither America, nor any single other country, owns the truth. We have a responsibility to look beyond ourselves.

Leaders and Followers

In George Orwell's *Animal Farm*, the animals on an English farm revolt successfully against the self-serving management of the human farmer and start running the farm for themselves.[7] The book was intended as a satire of the Russian revolution and its aftermath. As in any human revolution, Orwell's animal version rapidly deteriorates into another self-serving system in which the leaders (in this case, the pigs), helped by some hangers-on, exploit the gullible followers. Orwell's evocation of the social developments among the animals and of the personality traits of the various key players in the story is an uncannily perceptive, though somewhat cynical, metaphor for human society. Even in sophisticated modern democratic societies with high levels of education, we recognize certain people as born leaders. We are willing to grant such individuals positions of power over us, and we expect them to use that power to increase their social status and wealth. In democratic societies, we place certain limits on our leaders' power and on their freedom to manipulate and exploit others. In dictatorial or other hierarchically structured societies, such limits are often very ambiguous, or even absent. Why are so many people willing to accept the clearly selfish behavior of their leaders, allowing them to enrich themselves at society's expense?

There is no simple answer to this question. In most cases the leader gains power first in a definable section of society such as a gang, a political party or the army. Then, with the help of this sub-group, he begins to control larger sections of society. Such leaders are often able to remain in power indefinitely, as long as they maintain the support of the gang of henchmen that promoted them to the top. They usually lock in support by feeding the greed of these gang members, who in turn manipulate and terrorize citizens into compliance. Josef Stalin succeeded with the help of the Communist Party and the KGB, Saddam Hussein did it with the Baa'th Party and the Revolutionary Guard, the Roman emperors did it, and Orwell's pigs did it.

Even our ancestors, from long before we became human, probably used the same tactics. Some chimpanzee alpha-males are maintained in a position of power by a political alliance of other members of the community, who for selfish reasons prefer the current alpha-male to potential alternatives. The most primitive human societies also secure leadership through this system of a powerful individual supported by loyal underlings. Among such tribes as the horticultural Yanomamö, the village leaders acquire and maintain their position by a combination of simply being recognized by the majority as the best person to lead the group and by depending on the support of their patrilineal-kin group. As we saw in the previous chapter, the leaders of such groups, and to a lesser extent, their henchmen, tend to have greater wealth, leisure and access to females than the average male in the group and leave more viable offspring.[8]

Among the Yanomamö and other societies that are subdivided into small sovereign units, as well as in modern democracies, leadership is maintained more through psychological manipulation than brute force. Sophisticated political maneuvering, bribery with material rewards and real, or symbolic, titles allow the leaders to keep a loyal gang of henchmen behind them, who in turn manipulate the public at large with promises, entitlements, and with the most powerful weapon: fear of real or perceived enemies.

Had our leaders through the ages merely wanted a somewhat more affluent status and the means to manipulate the economy so as to reward their supporters and penalize their opponents, little harm may have come to the general populace. Many leaders of small western democracies are, and have been, of this ilk. These men or women often do a commendable job while in power and are revered long after voluntarily ceding their temporary leadership role. Some of the leaders of primitive societies described by anthropologists are also such people. Unfortunately for mankind, another variety of leader is not satisfied with such modest rewards and plays dangerous games between sovereign societies. These leaders seek to enrich themselves and consolidate their power by taking their societies into unnecessary wars.

The real question is not why such individuals seek extremes of wealth and power, but rather, why do human beings so willingly trust untrustworthy leaders? To understand why we are so easily hoodwinked, we must once again take a look at the evolution of our species. Long ago, the willingness to be led and to make sacrifices for a perceived common good was deeply engraved in the human psyche.

The Distant Origins of War

In our closest relative, the chimpanzee, some young apes develop into leaders, while other males show no aptitude for that role. These consistently observed distinctions in status indicate that in all probability our ancestors structured similar community hierarchies for at least five to ten million years. Individuals play other interesting roles in the chimpanzee political structure. One role is that of the henchman, either a single individual or a few individuals functioning as an alliance, which actively supports the alpha male. A variant of this role is a henchman, or an alliance of henchmen, who support a rival to the leader. On occasion, this rival faction succeeds in overthrowing and demoting the alpha male. A particularly clever role player may pretend to be a faithful henchman, only to challenge the leader after building up a following of his own. As primatologists study our ape relatives more closely, they are discovering just how sophisticated ape politics are—although, both the lack of a complex language and the limited size of the ape community place severe limits on the development of either a structured despotic government or consensus-based democracy.

If we are right in considering the leader-to-follower continuum to be essentially genetically based, we have to explain the general observation that alpha males tend to be reproductively much more successful than the average male, yet born leaders are relatively rare in both ape and human societies. Through selection of leadership genes, would not the higher reproductive success of leaders have resulted in populations of mostly leaders? The explanation probably lies in that interesting form of natural selection called *frequency dependent selection*, which we have already seen as a possible explanation for the maintenance of extreme socio-psychopathic behavior in human society. The genes that are necessary to make an individual into a born leader will be strongly selected in a population in which such genes are relatively rare, but as they gain in frequency, too many aspiring young leadership candidates enter the arena, and other kinds of selection start to limit the advantages of being a born leader. Fighting will lead to injury and death for some. And others may be socially punished while still young or shunned by females later in life. Finally, leadership candidates often seek to prove themselves by participating actively in raiding of neighboring communities. This can pay big dividends—or be a death warrant.

Raiding in chimpanzee society among adjacent communities is deadly serious business. The primatologist, Anne Pusey, has described the belligerent activities during a war between two chimpanzee communities as follows: In the Gombe research area in Tanzania, the males from one community, the Kasekela, systematically raided their neighbors, the Kahama, until they had killed all of the males, causing the permanent collapse of that community. The remaining Kahama females died or dispersed, joining other communities in the region at the bottom of the social rank order. The larger communities in such wars take over the feeding territories of the defeated smaller communities, and if successful, can eventually split into two independent communities.[9] Raiding at this intensity is essentially civil war between two neighboring, genetically-related sovereign communities of the same species. This more structured war, among internally relatively peaceful chimpanzee communities, has evolved independently from the similar situation in the African lion, where prides fight one another along the borders of their hunting territories, often with lethal results.

We may wonder why the chimpanzees could not peacefully share their feeding territories, thereby avoiding all this fighting and killing. The simple answer: they cannot. The productive limits of the forest, to which chimpanzees are adapted, determine the maximal sustainable healthy chimpanzee population density. We cannot expect chimpanzees to practice birth control, nor can we expect them to prefer starvation and disease within their community as a form of population control. Having out-evolved their traditional predators, war among communities is their best strategy. During periods of resource shortages, stronger communities will wage war on weaker neighbors, forcing them into defensive conflict. The conclusion we must accept is that in the world of the chimpanzee war is inevitable. But, we are not chimpanzees; we are Homo sapiens, who have the ability to reason and to reflect on past experience.

Early Negotiations for War or Peace

The chimpanzee is not a recent ancestor of mankind. However, some six million years ago, we did have a common ancestor from whom both the chimpanzee and our species have descended. This forest ape in all probability was primarily a forager, but it also hunted. It was polygynous and probably lived in communities very much like the chimpanzee does today. From such an animal, we evolved step-by-step into hunter-gatherers who, by some twenty thousand years ago, were fully human.

Throughout early human evolution, hunter-gathering groups gradually changed from living under the domination of selfish and despotic alpha males, whose domination depended primarily on physical force, to living under the leadership of headmen who used argument, example and psychological manipulation to lead their groups. At the same time, the ability to use language for argument and psychological manipulation opened the door to the possibility either to plan and assess more complex acts of war or to negotiate peace. We again mention the well-studied horticultural Yanomamö of the upper Orinoco rainforest. Their political system is very sophisticated, even Machiavellian in its emphasis on intrigue, deception, conspiracy, trickery and subterfuge. The complex within-village patrilineal kinship rules and status hierarchies spill over through between-village kinship relations and mistrust into a constantly shifting pattern of alliances, economic domination and overt hostilities.[10]

It is a serious small-scale and relatively simple version of typical relations among sovereign human communities, yet the difference between the Yanomamö and the chimpanzee situations is enormous. What happened during the eons of early human evolution after we split from the chimpanzee lineage and before the development of agriculture? If we want to speculate how our ancestors lived, competed, fought and survived, we can look at the fossil record and at nineteenth and twentieth century studies of the last remaining pure hunter-gatherer communities.

What the Fossils Tell Us

Roughly six million years ago, the common ancestor of the three now extant species—chimpanzee, bonobo, and man—split into two lineages. One, the genus *Pan*, stayed in the rainforest, while the other moved into more open savannah woodland habitat, and eventually became us, *Homo sapiens*. For the first four million years of our development from ape to human, we were anatomically still ape-like creatures classified broadly into the genus *Australopithecus*. Early during this period, we became effectively two-legged hunter-gatherers, though our shoulders and limbs remained structured for climbing trees as well. Most significantly, the fossils show that throughout this period, males were thirty to fifty percent larger than females. This indicates that our ancestral *Australopithecus* species was highly polygynous, with males adapted to fighting. More than likely, their aggressive behavior included not just within-community fighting, but also lethal war-like fighting. We can conclude this from studies showing that all highly social animals "externalize" violence from within-the-group to between-groups.

Approximately two million years ago, one branch of the *Australopithecus* genus, *Australopithecus habilis*, started to show some changes, including larger brains and more varied stone tools. In fact the changes became so extensive that most experts consider this the time that the species *Homo erectus* was born.[11]

Compared to its ape-like ancestor, *Homo erectus* was no longer a tree climber, had a bigger body, longer legs and a much larger brain. Most importantly, males were only slightly larger than females, much the same ratio as that of modern humans. What does this suggest about the evolving species' propensity to wage war? As noted above, the greater male-female size ratio of our early ancestors was probably the result of a polygynous society, with males doing a lot of fighting. Does the relatively smaller male of *Homo erectus* mean that the males became more peaceful? Probably not. We know that *Homo erectus* mastered new technological skills; fire became a manageable asset, some foods were cooked, and tools and weapons were much improved. As the Biblical story of David and Goliath tells us, an effective weapon and the skill to use it count for more than size and brute force.

In his book, *La Guerre du Feu*, the Belgian novelist J. H. Rosny speculates that mass murder inflicted by man upon man could have been perpetrated by hunter-gatherer groups simply by putting out another group's fire in an Ice Age winter. We do not know this as fact, but we do know that fire from volcanoes and wildfires was used before man learned to start fire. During those early days, fire was a precious, life-enhancing resource, one sometimes worth killing for. Rosny explores how early human groups possibly engaged in deadly competitions to monopolize fire just as nations today try to monopolize and control nuclear weapons.[12]

At every stage of our evolution, where we have evidence of how our ancestors lived and how their bodies were adapted, there are indications that we come from a long line of warriors. It is also true that from *Homo erectus* onwards, the fossil record tells us relatively little about whether we lived at war or in peace. The sad truth of what condemned the chimpanzee to perpetual warfare holds equally for our hunter-gatherer ancestors. For most of their existence they had good defenses against predators. They would have preferred to fight wars against neighbors rather than starve in the desert or fight among themselves, and planned population control would have been adaptive only for groups living in very unique environments, such as the Purari people in Chapter Four.

Does this mean that mankind is also permanently condemned to wage war? We do not think so. Yet, we believe that hunter-gatherers must have been warriors for the very same reasons that chimpanzees are—to survive in an environment with limited resources. To test this prediction, let us look at the evidence that archeology and anthropology provide about the lives of hunter-gatherers.

What the Evidence Tells Us

Before the development of stone projectile points, some thirty-five thousand years ago, there is very little skeletal evidence that can pinpoint whether a person died of violent trauma or of a non-violent cause. The closest we can come to designating skeletal remains as victims of war before forty thousand years ago is with the very high percentage of Neanderthal skulls that show evidence of serious head injuries. Because of the sheer number of damaged skulls, it is hard to imagine that all the injuries were caused by accidents or were the result of individual acts of homicide.

For the last thirty-five thousand years, the evidence for war before the advent of agriculture is much more convincing. In several parts of the world, both single and mass graves have been found containing skeletons in close association with stone projectile points, some of which were partly embedded in the bones of victims. For example, at one late Paleolithic site (approximately thirteen thousand years old) near Jebel Sahaba in the Sudan, fifty-nine skeletons of men, women and children were found, most exhibiting evidence of war injuries. All the children had been shot by arrows in the head or neck, indicating their execution as the last stage of a genocidal battle. From the Mesolithic and Neolithic periods up to the beginnings of agriculture, there is far more archeological evidence of hunter-gatherer warfare, including cave paintings of battles between archers. The quantity of this type of evidence is a strong indication that war was a frequent event in prehistoric hunter-gatherer societies.[13]

Not surprisingly, historical records also show most hunter-gatherer societies studied in the nineteenth and twentieth centuries tended to wage war frequently, either aggressively to capture resources, such as waterholes or hunting grounds, or defensively to maintain control over these means of survival. Australian aboriginal foragers, the Kung San of the Kalahari, the Copper Inuit of the Canadian Arctic, all of the West coast fishing communities and many other hunter-gatherer tribes throughout the world practiced this pattern of warfare among neighboring, sovereign communities. There are, however, some interesting exceptions. These usually are descriptions of people, living under exceedingly challenging conditions, who have high mortality rates and low birth rates not caused directly by resource limitation. Such communities have often been defeated by previous neighbors, after which they fled into their current inhospitable areas. Anthropologist Lawrence Keeley, in *War Before Civilization*, presents a very convincing case for war among hunter-gatherers as a normal component of social behavior[14]. He notes that our ancestors for at least the past six million years have lived under conditions in which their communities were virtually always under threat of attack. Attacking one's neighbors first must often have been seen as the best survival tactic.

Escalating Aggression

Once people began to grow and harvest their own food, the face of warfare changed. Agriculture, even the most primitive kind, implies an attachment to a specific bit of land, the accumulation of equipment and other materials and a sedentary life moored by some form of at least semi-permanent dwellings. In many cases, the development of agriculture also led to the establishment of larger communities, such as villages. Under these circumstances, abandoning one's home became much more costly, which meant that more effort was spent on defense. History is filled with examples of entire tribes and nations fleeing from the aggression of stronger neighbors.

On the face of it, one would not think that the shift from foraging to growing crops would significantly affect warfare. But higher production of food per acre in horticulture or agriculture, compared to simply harvesting what nature provides, had two destabilizing effects. First, it allowed populations to increase until they reached a new maximum, determined by the maximum food production capacity of the managed land. Improved methods and new crop varieties could postpone the crunch, but sooner or later only the acquisition of more land could ensure the survival of a growing population. The second destabilizing effect of agriculture was that every now and then a major crop would fail, or the quality of the land would slowly deteriorate, resulting in gradual loss of productivity. While hunter-gatherer communities also faced food shortages, they typically affected relatively small numbers of people. Agricultural communities had much larger populations as well as much better weapons. Hence, failed crops created large, armed congregations of hungry, passionate people on the move. These people often convinced themselves that their crops failed because the locusts or drought were inflicted upon them by their evil neighbors' sorcery. They would have vilified them as the "enemy" and eventually waged war against them.

The larger population size of agricultural communities also led to a greater division of labor, and the consequent complex social structures demanded a stronger leadership in the form of headmen, shamans, elders, oracle priestesses and other individuals of influence. When their leadership failed to deliver anticipated benefits, they would have blamed the unpredicted bad times on others. Religious leaders would have consulted their gods or goddesses and claimed that these deities identified their neighbors as the source of evil. Humans have always tended to form belief systems that would allow them to blame misfortune of any kind on others.[15] By the same token, we have also given credit to illusory leaders for helping us conquer an enemy. The Israeli general Moshe Dayan was reported as having claimed that during the Six Day War, Israeli forces were fighting on God's side. When we believe that any god or goddess is "the general," victory and defeat in wars can be blamed on powers beyond ourselves. This is a dangerous perception and one that could eventually cause our annihilation.

Whatever we, as individuals, believe about war, it is crucial to understand that it has been a part of hominid biology and psychology since the beginning of what we know as our history. It seems to be an expression of one of the darker aspects of our nature. If we are ever to build cultural structures that will guide us towards a more peaceable world, we must recognize and hold in check those parts of our psyches which compel us towards violence—towards war.

War within Our Psyches

History has screamed from the Promethean discovery to the present day that many humans desire what we perceive to be life enhancing and believe we have a right to take it through any means available. We may have inherited the darker shadows of greed and jealousy from ancient ancestors, but hopefully our ability to make choices that could transform our destiny is also an evolved component of our biological heritage. The most critical question is why have we not yet learned that violence will only result in more violence and that negotiating for cooperation resolves conflicts with less destruction to the world than deadly competition?

Author H.G. Wells shed some light on this question when he wrote "Moral indignation is jealousy with a halo."[16] Moral indignation puts blinders over the eyes of our souls, allowing us to covet what does not belong to us, rationalize that we deserve to have what others have and manipulate them or use force to possess what we want. Once moral indignation looms large, we are prone to call in greed, jealousy and self-righteousness to support it.

Self-righteousness is most obvious in civil and religious wars, which are usually fought over perceived moral and ethical disagreements. It's as if we are compelled to believe that "we" are right and the other wrong. One of the strangest character quirks in human psychology is that we often prefer being right to being happy. Our psyches seem to delight in the validation of being right, regardless of the price paid to prove a point. In war, the point is often forgotten and the only important goal is to win—to kill more of them than they kill of us. War uncovers and lays bare our capacity to act in barbaric, inhuman and destructive ways. It calls forth the darkest and most inhumane aspects of our nature.

Following the Civil War in the "Dis-united" States, General Robert E. Lee made the comment: "It is well that war is so terrible, or we should grow too fond of it."[17]

In the early stages, war does energize a country. Excitement and nobility often emerge when a "call to arms" is being heralded across the media. Fear of being attacked by some proclaimed enemy is somewhat diminished when we believe our powerful leaders and military will save us. War correspondent, Chris Hedges, writes in *War is a Force that Gives Us Meaning* that when soldiers rush into a battle, they often report feeling excitement and euphoria, kin to an addict taking a hit of his drug of choice. Fighting for one's country gives some an opportunity to gain status, to become a hero, to be seen as an important member of a group, which gives powerful meaning to their existence. During his early years on battlefields, Hedges writes that he would have rather died than go back to a mundane, routine life. The battlefields offered intense and overpowering moments of passion. Even if it demanded his life, he felt it would be worth it to have experienced that degree of passionate aliveness. But after years of observing the suffering, carnage and evil that is prevalent in war, he came to realize that what war actually does is expose the evil that lurks below the surface in the every human psyche.[18]

One of the most damaging aspects of a call to arms is that our perception of reality becomes skewed. When we begin to perceive "the others" as evil and we separate "us" from "them," we deny important aspects of ourselves, including our personal evil side.

In the final analysis, we all are more alike than different. We are all connected and interconnected, dependent and interdependent. We need the same resources to preserve life. We must learn to share one planet, if we are to survive. The bottom line truth of war is that nobody wins.

References:

Q. Bradley, James. *Flyboys: A True Story of Courage.* New York: Little, Brown and Company, 2003.

1. Bradley, James. Ibid.

2. Keen, Sam. *Faces of the Enemy: Reflections of the Hostile Imagination.* First published New York: Harper and Row, 1988. Revised, expanded Edition, 2004.

3. RAF Bomber Command Diary.

4. Bradley, James, *Flyboys: A True Story of Courage.* New York: Little, Brown and Company, 2003.

5. Internet, report. info@votevets.org. 2002

6. Byrd, Robert, U. S. Senate Floor remarks, May, 2003.

7. Orwell, George. *Animal Farm.* London: Secker and Warburg, 1945.

8. Chagnon, Napoleon. *Yanomamö: The Fierce People.* Fort Worth, Texas: Harcourt, Brace Inc., 1997.

9. Pusey, Anne E. *Of Genes and Apes: Chimpanzee Social Organization and Reproduction.* In de Waal, F. B. M. (Ed)—2001—*Tree of Origin – What Primate Behavior Can Tell Us about Human Social Evolution.* Cambridge, Mass.: Harvard University Press, 2001.

10. Chagnon, Napoleon. *Yanomamö: The Fierce People.* Fort Worth, Texas: Harcourt, Brace Inc., 1997.

11. Wrangham, Richard. "The evolution of sexuality in chimpanzees and bonobos." *Human Nature,* 4(1), 47–79. 1993.

12. Rosny, J. H. *La Guerre du Feu.* In Wright, Richard, *A Short History of Progress.*

13. Wendorf, Fred. *The Prehistory of Nubia.* Vol. 2. Dallas, Texas: Southern Methodist University Press, 1968.

14. Keeley, Lawrence H. *War Before Civilization.* Oxford, England: Oxford University Press, 1996.

15. Watkins, Trevor. *The Beginnings of Warfare.* In Hackett, J. (Ed.) *Warfare in the Ancient World.* New York: Facts on File Publishing, 1998.

16. Wells, H. G. In Bartlett's Quotations.

17. Lee, Robert E . In Bloom, *The Lucifer Principal.*

18. Hedges, Chris. *War is a Force that Gives Us Meaning.* New York: Public Affairs Press of Perseus Books Group, 2002.

CHAPTER

Homo Libidinosus

The most significant battles are waged within the self—over love.
~ Sheldon Kopp

Walking along the shores of Omaha Beach, following the invasion of Normandy by the Allied forces on June 7, 1944, revered war correspondent Ernie Pyle reported devastation so severe he had no words to describe his reaction; he could only write exactly what he saw. He could see men sleeping on the sand—many who would never wake up. He could see the personal never-to-be-used-again gear of the soldiers who fought and died to give the Allies a toehold in Europe. This gear, strewn for miles along the beach, included socks, sewing kits, diaries, Bibles and hand grenades. There were also letters from loved ones and photos of sweethearts, wives, children and parents, all staring up at him from the sand. Pyle picked up a Bible, perhaps seeking words that might help him understand what surrounded him. Later, he dropped it back into the sand. As he did this, he noticed a young, dying soldier, who held a large, smooth rock in his arms, as if it were his last link to a vanishing world. He was softly crying, "Mama, mama."[1]

Many war correspondents have reported hearing dying young soldiers cry for their mothers. More often than not, our need for love seems to surface in our darkest hours. It may well be the only thing that can hold the dark at bay. Just as journalist Ernie Pyle could not describe his emotions at the overwhelming devastation around him, enduring love is beyond the reach of language; it can only be understood through experience. Economist Adam Smith showed keen insight when he stated, "The chief part of human happiness arises from the consciousness of loving and of being loved."[2]

What is Love?

A sage friend, a husband, a father and a grandfather, gave me pause recently when he asked, "What is love anyway?" I (Paddy) have written two books on the subject, which he'd read, but he didn't want a "book answer." He wanted a response that came from my heart and soul. Reluctant to attempt an answer that might miss the mark, I changed the topic to something safer, but the question has reverberated in my mind ever since.

We use the word "love" to express our feelings about anything from enjoyment of our favorite ice cream to our deepest emotional commitments. Perhaps we should have many different words representing the various kinds of love, as we are told the Eskimos do for snow. In trying to categorize the ways most humans refer to love, we recognize a few distinct categories of love. Each category involves a connection, or attachment, to something or someone that elicits desire to protect, cherish, possess, or otherwise keep the love object in our lives. Because to love is an internal, subjective, emotional response, we can only love to the extent we are willing to be open to our emotions and emotional needs.

The simplest kind of love, related to our enjoyment of something—such as a certain food, a piece of music, a sport or activity, might be called "irreciprocal" love. Irreciprocal love is essentially a one-way street, since whatever we love does not possess the level of consciousness required to love us in return. This is not the same as unrequited love, in which a conscious choice is made not to love in return. Irreciprocal loves are often fairly casual loves, such as our desire for a dish of fresh strawberries in early summer, but can also be deep-rooted loves for essential resources and conditions. Casual irreciprocal loves usually come and go without causing any major difficulty in our lives. However, these loves cease to be casual when they offer us more pleasure than our more complicated relationships with other human beings. The more comfort, pleasure or excitement offered, the more attached—sometimes, even addicted—we become to whatever the object of our irreciprocal affection might be.

For example, I (Paddy) would never issue my husband, Tim, an ultimatum to choose between his glider (soaring plane) and me. He would offer to help me pack my bags within seconds. What may be a casual irreciprocal love for some of us can involve deep feelings of passion and commitment for someone else. Although I also fly gliders and enjoy soaring, my heart would not break if I were told I could not fly again. Tim's might. The distinguishing characteristic of this category of love is that whatever the object of pleasure, it does not have feelings for us. We can become passionately addicted to whatever we love in this category, but it would not cry if the connection were broken. For some people, this may be an advantage of irreciprocal love; it is free from the complexities of mutual feelings and responsibilities.

The other categories of love involve connections to other sentient beings. The umbrella term for these connections is "relationship love." Relationship loves can extend to and be shared among people connected through groups, such as neighborhoods, tribes, communities, religious sects and nations. Sometimes, relationship love is experienced with pets, especially dogs. These relationships can offer us pleasure and some degree of mutuality. Relationship loves can often be challenging. This is not a strike against loving, just a reality of its nature—of our nature. It does not take us long to learn that loving others causes us to feel vulnerable, and therefore, we are open to possible disappointment and the risk of emotional pain. This pain may not be caused intentionally, but happens due to the dynamics of shifting needs and interactions, as well as the natural cycles of life and death. The meaning of relationship love is very personalized for each of us, as it is rooted in our deepest personal experiences. It often seems that the longer we live, the more confused we become about what relationship love could, or should, embrace.

Individual relationship love can be divided into many different categories, such as parent-offspring, sibling-sibling, extended kinship, friendship and the most complex— partnerships that involve sexual commitments, both short-term and stable, enduring ones. Some of our relationships may be on casual terms; however, we are never able to control everything that might transpire between ourselves and another sentient being, as their emotions and behaviors will also impact the relationship. This fact alone causes relationship love to be more difficult than loving something without a developed consciousness. In addition, we each have some unfinished business, unresolved pains and unmet needs in the love department, which causes relationship loves to be more complex.

We may have fairly successfully moved through the three major stages of love—attachment, empathy and compassion, but there will be some wounds that continue to bleed into future relationships.

The Origins of Love

Attempting to find a scientific description of love and an explanation of how it evolved into the complex set of emotions humans experience is a daunting task. Many people may believe that it is inappropriate to even make an attempt, since our own emotions regarding love will interfere with objectivity and rationality. Yet, as scientists who realize that our relationship behavior is a major determining factor of our survival, we believe we have a responsibility to try. Hence, we will again step back and take a look at some animal species other than our own. In Chapter Two, we touched on the evolution of emotion and how it preceded reason in the tool-kit of life. Love might well have its evolutionary beginnings as the emotion an animal feels when experiencing or imagining conditions, activities, events or objects that enhance its fitness—that is, its likelihood of survival. This will serve as a rudimentary working definition of love: an attraction to that which we believe promotes our survival. When someone asks if you would enjoy a summer picnic, your brain will immediately do a pop-up of images, real or imagined, of sunshine on water, of good food, of friends and much else, and you might reply that you "love" summer picnics. It seems that non-human higher vertebrates experience similar emotions when facing or imagining conditions that are enjoyable for them. Evolution has matched love-responses to what is fitness-enhancing.

Polar bears love a variety of things, conditions and events. They love fresh and putrid seal meat, berries, a swim on a hot day and a snooze in the willows. A male polar bear loves mating with female bears in heat. A female loves her cubs and cubs love their mothers. It is not difficult to recognize that everything a polar bear loves can be placed into three categories: material aspects of its environment, sexual experiences, and close relatives. Each of these categories serves the bear's fitness in a different way, through natural selection, sexual selection, and kin selection respectively. We can therefore assume that the bear has evolved three different kinds of love in response to these selective pressures. While we cannot know what the bear actually feels, we do know how the bear reacts behaviorally and physiologically to the different kinds of love-eliciting experiences. It will come as no surprise that the intensity of the bear's irreciprocal loves varies with the importance of the stimuli and his needs at the time, but the bear is rarely willing to risk his life for whatever it loves in the material environment.

However, a male bear will fight fiercely for access to a female in heat, and a sow will do just about anything to protect her cubs. The stronger the fitness-enhancing stimulus, the greater the effort and risk a bear will invest in this stimulus. We assume that greater effort and risk taking means that such actions are driven by stronger love emotions.

Polar bears are not social animals, which means that they have a rather simple emotional life. The more complex interactions of natural, sexual and kin selection, as found in social species, such as the muskox, pigs and geese, are absent. The most complex of bear loves are those shared between a sow and her cubs. Parent-offspring relations are very old in an evolutionary sense and have both material-love and kin-love components. The long history of this special kind of belonging love, which we humans share with many species, is of great interest to us, as it seems to be the springboard from which life partnership love sprung.

Life Partnership Love

Belonging love took a big step forward when relationships began to involve two specific unrelated individuals. One of the oldest and most intriguing developments in this direction is long-term pair bonding in animals such as geese. My wife and I (Dolf) once kept a pair of geese on our farm. Lawrence and Margaret were an inseparable pair. In spring, when Margaret was receptive, she and Lawrence would court and mate frequently, resulting in Margaret laying a nest full of fertile eggs. It had to be the right season for Lawrence to be sexually ready for action and Margaret had to be in the right mood so to speak. Her responsive mood began with the lengthening of the days in spring, and ended after she had laid about a dozen eggs, which was the end of sex for the rest of the year—but not the end of their relationship. These two geese had bonded for life. They stuck together at all times—preening one another, feeding peacefully together and sleeping together. The strength of their love became even more evident to us when a neighbor's dog killed Lawrence. Margaret called for him for days. She refused food, lost weight and sat in a corner of the barn, shrinking into a pathetic dirty ball of feathers. Fearful that she would die, we bought a young male goose, who promptly fell in love with her. In response, Margaret attacked him viciously and drove him away. After persistent overtures by the male over the next couple of weeks, she finally accepted him, but she was never the same as she had been with Lawrence. She did not lay eggs the next spring and died the following winter.

Most animals that form pair-bonds do so only for the duration of a single breeding season. If the bond is for succeeding breeding seasons as well, the parents do not necessarily stay together outside of the breeding seasons. This is the case for albatrosses and many other species of sea birds. Forming a pair-bond can enhance fitness in several ways for both sexes, adding up to the increased survival of their joint offspring. But what is the advantage for staying together for a lifetime? And why did Margaret show such intense and lasting grief after losing her mate? The answer lies in the evolution of joint learning of cooperative survival traits, of learned dependency on one another and of the rewards that they experience as a result. Margaret and Lawrence were not genetically related; in fact they were of two different breeds of tame geese. Their love for one another was not based on kinship, and sexual love was only a small part of their relationship. Their love grew after they were brought together as yearlings and learned from one another how best to survive in their barnyard environment. Learning new behaviors, especially from a partner, and learning the benefits of acquiring such traits are a part of fundamental culture. It is generally accepted that higher primates have culture, but we argue here that geese and many other birds and mammals also have enough culture to have influenced their evolution. Geese have evolved a kind of love emotion and its associated behaviors similar to the behaviors that we humans experience in enduring love relationships. It works for geese for the same reason it works for us.

Human pair-bonding is a fascinating phenomenon because it is a relatively recently-evolved major change in how we organize our social and reproductive lives. None of the other great apes pair-bond as part of their social structure. Our closest relative, the chimpanzee, is a classically promiscuous species, although it has been reported that a female in heat and a male will sometimes separate from the other community members and spend several days alone in a brief monogamous sexual relationship. There is no reported evidence of anything like "falling in love" in apes or other non-human primate species. Where does this passionate and sometimes destructive human emotion come from? Having developed a more all-inclusive and more complex culture than any other species, and being a more or less monogamous breeder with extremely slowly developing young, it would seem reasonable to search for an explanation for this uniquely human phenomenon. To help us understand its origins, we will delve into the first kind of love we experience. C. S. Lewis claims that we have to be loved to survive our beginnings and must continue to love to stave off existential loneliness.[3]

Parent—Offspring Love

The love a parent feels and expresses for her child may be the oldest love in the world, but the intensity, complexity and duration to which it has developed in our species is unique. Its matching counterpart, the love a child develops for his parents, is different in both origin and nature. What is most exciting about these two loves is their interplay, and how the child learns through this interplay what love is and how to make it a driving part of life. The two components of this interplay, the parent's love and the child's love, have co-evolved; each is a component of the natural selection on the other. One's first experience of love is kindled within our psyches with our first sense of relationship attachment, of belonging to someone who will help us survive. It began for me (Paddy) when my mother first offered me nourishment from her body. It grew through the nights she rocked me to sleep, singing *"Hush lil' baby, don't say a word, Daddy's gonna buy you a mocking bird."* As I replay that tune in my mind, I am flooded with memories: Mother's staying up into the wee hours of the morning making doll clothes for the doll I would be getting for Christmas, with love in every stitch. She did not enjoy sewing; in fact, she often commented she hated it, but she did it because she loved me. I remember the special snacks she always had ready when my brother and I came home from school and the delicious meals she prepared to accommodate our schedules. I was blessed because my parents truly loved each other throughout the 57 years they shared. We never heard them argue. I'm sure they did, but never in front of the children. She was a mother who believed in fulfilling the duties and commitments of parenting. She always made birthdays and holidays special. Sunday dinners were a weekly feast, and she ate the chicken wings, insisting they were her favorite. She didn't often say, "I love you," but she lived it—and I felt loved.

Before we are born, nestled in the uterine home of our birth mothers, we are connected to a supply of all the nutrients we need to develop and grow. Some psychologists believe that this primal connection represents a kind of attachment that we try—in vain—to re-establish for the rest of our lives. As infants and young children, we love out of our need to belong to someone who will take good-enough care of us to allow us to survive. Therefore, attachment, defined as a sense of dependency, can be seen as the first stage of belonging love. Our primary caregivers are our first love teachers, and we will do anything to maintain this connection. The ways we are treated by them—including how we are touched, looked at, talked to and attended to—become automatically our first definition of love, even if some of their behaviors are not loving. As Sam Keen says:

> *If your mother burned the toast, then love means burned toast!*
> *It is as if the ways they treat us and speak to us forms a grid*
> *in our minds by which we humans begin to experience love.*[4]

Those of us who experienced safe attachment and were well-cared for during infancy and early childhood have an emotional advantage when it comes to moving from the primary attachment stage of love to the next stage, empathy. Empathy is the ability to feel what another sentient being may be feeling—to walk in another's moccasins, so to speak. Attachment is about ME feeling loved. Empathy is about I-Thou, the relationship exchange between ME and YOU. You begin to matter to me and can influence my feelings by your feelings. I become concerned about how you feel and not just how you are treating me. However, the critical prerequisite of empathy is having felt loved enough in early child-hood that we feel secure in sharing it with another. Love usually engenders more love.

The notion that we have our first love lessons in infancy has been confirmed through behavioral experiments with non-human primates. In the early 1950s, research psychologist Harry Harlow carried out a number of now-famous studies with rhesus monkeys to determine if the infant's later-life behavior, including expressions of love, depended on early interactions between mother and infant. He found that infants who were reared by mechanical surrogate mothers, deprived of soft furry material to cling to, were more fearful, nervous and aggressive than monkeys raised by either real mothers or surrogate mothers, made of soft, furry cloth. The infants reared in the absence of contact with real or soft surrogate mothers, when fully mature, did not develop normal, effective sexual behavior. When these deprived females were artificially impregnated, they showed no interest in caring for their offspring.[5]

Since we cannot morally or ethically do designed experiments of this nature on human subjects, there is no solid scientific evidence that we humans also need loving parental care early in life in order to develop into loving, caring adults. However, a vast body of well-recorded anecdotal evidence supports this idea. When an infant's needs are not attended to in a predictable manner and the infant remains in doubt about his attachment to another being, the infant does not continue to send out distress signals, but becomes passive and depressed. He essentially gives up on love. It becomes more difficult, if not impossible, to jump-start love at a later stage of life.[6]

Sexual Love

As we age out of childhood, our biological and psychological needs shift, triggering a new intricately interwoven set of needs—not only to be acknowledged and loved, but also to have sex. Again, evolution influences our drives and choices. We become attracted to those who could help us reproduce, casting our genes and the traits they determine into future generations. For most animal species, sexual love is short-lived, simple and passionate. Since sex is so closely linked with reproduction, it is not surprising that the passion for sex is one of the most intense, as it is the fastest route to increasing fitness. Remember the passionate muskox bulls—no sex, no progeny; no progeny, no fitness. The muskox bull becomes sexually active in the rutting season, but even then, he only shows interest in a female when she is in heat. The bull needs two stimuli, the right season and the close-up scent of an ovulating female, to respond with overt sexual activity. We cannot be sure that the bulls experience love for their partners under such conditions, but having seen them in action, I (Dolf) would say passionate sexual enjoyment was definitely involved. It seemed obvious that the male loved sex, but whether or not he loved his sexual partner is another question.

In social animals, and especially in highly developed species, sexual love often becomes an integral part of maintaining social cohesion, apart from its potential for procreation. In our close relative, the bonobo, sex is commonly practiced as a form of bodily communication of goodwill and friendship among all categories of community members—mature, immature, male and female in any combination. By contrast, human societies place at least some restrictions on sexual activities, regarding who can have sex with whom, as well as where and when it is permissible. Such restrictions are not only enforced from the top down, but are usually experienced as "right" by the population at large. Those who break the rules are generally resented, often stigmatized, and, in some cultures, even prosecuted.

Since most human sexual activity does not lead to reproduction, it can be seen as a form of adult play that reinforces the love-bond between two individuals. Where we differ most clearly from the bonobo is that most of our sexual activity is restricted to pair-bonded couples—in either short-term or long-term relationships. While sexual love is an important part of pair-bonded love, it can also be expressed and enjoyed outside the pair-bond. In the best-selling book, *Men are from Mars and Women are from Venus*, John Gray asserts that women need love to be open to sex, but men need sex to be open to love.[7] While there may be a grain of truth to this assertion, it is a well-documented fact that both men and women benefit biologically and emotionally from a healthy sex life, whether in a pair-bond relationship or not.[8] There is sex for procreation and sex for recreation, but whatever the purpose, most adult humans enjoy sexual activity with consenting partners and are willing to put a great deal of time and energy into seeking sexual partners.

When reciprocal relationship love and sexual love coincide, we feel an intense caring, which allows us to bond deeply with another. This level of sexual relationship love relieves us of the anxiety of separateness. We can feel the ultimate sense of belonging, of being desired and of full aliveness. Because the deepest need of the human soul is to overcome the anxiety of separateness, our sexual needs compulsively drive us to form a union that can prevent our psyches from becoming depressed by emotional alienation—even if the union only lasts for a few moments. If we build an enduring and mutually satisfying emotional and sexual relationship with a lover, the security of knowing we are loved allows us to face our world with more confidence. When we are unable to maintain a relationship that meets our needs for affiliation—including sexual bonding—we become emotionally frustrated. Often this frustration becomes externalized into arguments, which, when unresolved, may cause aggressive and violent behavior (including rape) or can be internalized as depression.

Falling into Limerance

When we fall in love, we feel more fully alive. We can be almost overcome with passion and excitement when we fall under the spell of a new, potential love partner. The problem is that this phenomenon of "the fall" is based more on fantasy, hope and false expectations than on the reality of facing the challenges, opportunities and problems of sharing ourselves with another person. When the gap between the dream and reality is too wide to bridge, the experience often leads to the lover's falling into a state of limerance, or lost in an obsession of the idealized person. As discussed in Chapter Two, limerance makes it difficult to focus on any aspect of life other than the beloved person. When Dr. Tennov, who coined this term, studied the psychology of limerance, she discovered that this intensely passionate state usually did not involve sexual love, but rather the yearning for special attention from someone around whom a complicated fantasy had been constructed.[9]

Such was the case for me (Paddy) when I became obsessed with my eighth grade history teacher, Mr. Patrick. He was a real charmer and an excellent teacher, plus he had this way of looking directly into a student's eyes when answering her question. I used to lie awake until the wee hours planning my question so he would look at me "like that." Those looks made my heart flip-flop. Practically every girl in the class had a crush on him. We would get together and order pizza and beer to be delivered to his house and hope he would know how much we loved him. He would saunter into class the next day and say, with a grin, "Thank you, girls, but I would much prefer you spend your time study-ing and put your money in a savings account for college."

Then, one week, Mr. Patrick was ill, and Mrs. Patrick came as his substitute. We were aghast that there was an attractive, nice wife. She also thanked us for the pizza and beer, but explained that their children ate the pizza and they gave the beer to friends, because they preferred wine. By the end of that week, she had gained my respect and the reality of their life together had greatly diminished my obsession with my teacher. Still, over half a century later, I smile remembering Mr. Patrick.

Limerance is sometimes not so easily dismissed and often leaves a painful scar when reality sets in. But, regardless of the end result, our human hearts seem to latch on to people who bestow special attention upon us. Although falling in love can begin with many of the same dynamics as limerance, it is more likely to have a happier ending when friendship with the loved one develops as an important part of the relationship. In friendship, we put aside pretence and performance to build a genuine understanding of each other. If the friendship catches fire, and the passion is mutual, a sexually intimate love relationship may develop.

Soul Mates

The relationship shared by the geese, Margaret and Lawrence, embodied many of the qualities of deep friendship and love that many humans seek in long-term commitments. The most significant difference is that most mature humans need some private time alone to be able to happily share time with another, while Margaret and Lawrence were almost inseparable. Yet, these two geese represent in the animal world what we refer to in our species as soul mates, those who feel that they were destined to meet and to share some aspect of their lives. Soon after meeting, soul mates recognize their potential to relate emotionally, intellectually—and sexually, when appropriate. There is a special joy in being together that goes beyond the ordinary pleasure of other relationships. Soul mates rarely have to explain themselves to each other, as there is an intuitive understanding between them. They are first and foremost devoted friends. They may be genetically related—siblings, for example—or meet through happenstance. A dominant characteristic of these pairs is that when they are in groups, others sense a kind of magnetic attraction between them. Those who share their lives with someone they consider a soul mate are especially blessed.[10]

When I think of soul mates in my (Paddy's) own family, the first image is always of my maternal grandparents. My grandmother, Myrtle Dickens, was of the Charles Dickens clan and her family was considered "important" in their small town. Her mother died when she was ten years old, and her father remarried quickly hoping for help in raising his five children. Instead, the "wicked stepmother of fairy tales" became a reality for these children, especially for my grandmother. She effectively became

"Cinderella," expected to do the housework and look after her siblings while her stepmother attended social events. Myrtle's handsome prince showed up when she was fifteen years of age, but he did not invite her family to a masked ball. He was an eighteen-year-old nomad who had wandered into town from across the hills looking for work. He found a job—and my grandmother. They fell in love. Knowing their marriage would not be acceptable to her family, she tied the proverbial bed sheets together, climbed from a second-story window, and eloped with him. They raised five children and were together for fifty years. Shortly after their golden anniversary, as she was walking through a park close to their home, my grandmother was attacked by a large stray dog. Overwhelmed by fear, she had a fatal heart attack.

I will never forget seeing my grandfather afterward. His light had gone out. His body looked like a withered weed. His familiar chuckle seemed stuck irretrievably beyond our reach, and he told us plainly that he could not endure being alive without her. It was evident. His pain was visible in every aspect of his being. Even his hands, which had always been busy, were listless. Within a few years, he died of grief. At the time of his death, I felt a profound sadness that he'd never recovered from losing her. As I've matured, I've realized that they created and shared something most of us would die for—that rare and beautiful love in which two people totally entrust their hearts into the other's keeping.

Taking a walk with him several months after her funeral, I asked him what their secret had been. Looking up into the heavens, he replied, "She loved me *just as I am*. That's the most wonderful thing that can happen to anyone. Most people don't have any idea what love really is."

"What is it?" I persisted. At age twenty, I was beginning to understand that I probably didn't know.

It took him awhile to reply. "Well, you have to discover it with someone else helping you," he began. "It's like discovering that you are a king and live with a queen in a beautiful castle. We never had very much in the way our world puts values on things, but we really had everything—because we shared all we had."

That was over forty years ago. I doubt that I've ever heard a wiser definition of love. Love does not involve our possessions, but what we share of ourselves, trusting that it will be handled with care.

In serious conversation, there is near unanimity that the most rewarding and meaningful love is that between soul mates who bond for life—people like my grandparents. Yet, in popular literature, we thrill to stories of young couples who fall head over heels in love, while we commiserate with the brooding lover who suffers from a serious dose of limerance. Why do we so enthusiastically cheer the Romeos and Juliets of literature, while assiduously trying to prevent our own daughters from falling for the wrong man? We are probably subconsciously dealing with the reality that our species is under two conflicting selective regimes. One is our simple selection for those who produce the largest number of offspring. The other is the culturally driven selection based on social rewards and status awarded to families, such as the Dickens. For all its hazards, falling in love is a very effective first step on the route towards increasing one's reproductive output. We continue to evolve under the influence of many conflicting selective forces, often creating a society of psychologically fractured individuals, half ape, half god.

Enduring love is not uniquely human, nor does it always involve a sexual connection. Female chimpanzees frequently form close, enduring, loving friendships, and males who go raiding together or form political alliances also groom one another, expressing love for a comrade. Although soul mate relationships are rare in chimpanzees, they do exist. An example is the case of a three-year-old sickly orphan named Mel, whose mother had died. He had no one to help him cope with survival in the rugged hills of Gombe, Africa, and the members of Dr. Jane Goodall's research team were sure that he would die. Much to everyone's amazement, a twelve-year-old chimpanzee, Spindle, whose elderly mother had died in the same epidemic that had caused the death of Mel's mother, took a special interest in the young orphan. Spindle began to carry Mel on his back, just as a mother would have done. Spindle would reliably snatch his young charge from danger, even when the situation endangered his own life. In response, Mel showed dependent love for Spindle in the same way a child does for a caregiver.[11]

This example, and others from well-studied chimpanzee communities, indicates that the forming of close friendships involving overt loving behaviors in this species is based on astoundingly adaptable individual personalities. In the case of these two orphans, Spindle and Mel, recognizing the distress in the other, could have initially formed a mutual loving relationship based on empathy, but they became soul mates for life. This aspect of chimpanzee behavior and its complex emotional background gives us important clues about the kind of social behavior our ancestors must have had, which became the basis for our own even more complex emotional personalities.

Passion in the Garrison

Like love, especially when sex is involved, war calls us to the edge— that place where passionate emotions meld in a search for meaning. Although, love and war are imbued with their fair share of myths and fantasies, both promise rewards, if we endure the inherent battles. Each forces us to connect and to commit to something beyond ourselves, which gives us a heightened awareness of our own aliveness. Because passion includes the elements of excitement and vulnerability—even of the fear that is a part of vulnerability—it is understandable that war elicits deep passion. In 1688, the English restoration statesman, Lord Halifax, stated: "Love is a passion that hath friends in the garrison."[12]

Why does the word love come up so often when people talk or write about war? How can Chris Hedges say in *War is a Force that Gives Us Meaning* that even with its carnage and destruction, war arouses love and passion? There is no question that war does indeed give individuals, and whole nations, passionate feelings of belonging to something greater than ourselves. For some young people, especially those without a clear direction in life, war provides a cause, a reason and a way to contribute to something important. But such an attraction to war has a dark underside. If a love of war is deeply embedded in our hearts and genes, it points to a part of human nature that is totally incompatible with our very survival in a world with weapons of mass destruction.

When we watch the film "Lawrence of Arabia," most of us feel horror and shock at the scene in which Lawrence catches up with a group of fleeing enemies and goes on a killing spree. Afterwards, he admits that he killed because he enjoyed doing it and could not stop. In his personal journal, Lawrence wrote that he feared he had become addicted to the power of war and killing.[13] There are many other published examples of people who admit feeling the same—including the famed General George Patton, who led American troops in both world wars. He was often referred to as "Old Blood and Guts." Outside of the theatre of war, such killing is unacceptable to organized society, but on the battlefield it is the name of the game and society lauds the killers.

But there is a price to pay. Human beings have evolved to both kill and protect each other. These two drives collide in times of war, when we must kill fellow humans to protect other fellow humans. We know that through the use of propaganda, political leaders will try to instill enough hatred in us to make us willing to risk our own lives to kill the hated enemy. But when hating the enemy is not enough for us to overcome our powerful inhibition to kill, becoming caught up in the overpowering vortex of war mania can overcome this innate revulsion. One of the seldom-mentioned tragedies of war is that our evolved ability to kill leaves many warriors with an evolved guilt that haunts them for the rest of their lives. This deep psychological conflict is an inevitable result of our ecology. Just like the raiding chimpanzees, who are condemned by their ecology to frequent unending bouts of warfare, so are we, until we realize there may be more superior and cooperative ways to control our ecology. We will explore these alternatives in later chapters.

In the midst of our complicated, contradictory attraction to killing, another kind of significant relationship often develops among the warriors—a deep comradeship and love. When soldiers share the trenches and their lives become dependent upon each other, they frequently bond deeply and enduringly. Not only do they feel attached and empathic towards one another, but they may open their hearts to a deeper compassion for others. Compassion requires that we open our hearts and minds to accept that we all suffer and that we are all related. I have heard many veterans say that their enemies taught them as much about love as about hate, as was the case with the soldier in Chapter Three. Veterans frequently hold reunions that are filled with passionate love and passionate tears as they remember the ones who are not there. A WWII veteran, who had recently returned from a reunion of his Normandy invasion comrades, told me, "I've never laughed so hard, cried so openly, nor felt so loved as I did at this reunion." I (Paddy) felt tears spring to my own eyes as I felt his love for these men.

I asked how his feelings towards these men compared with his feelings for his wife, with whom I knew he shares a deeply loving and long relationship. Smiling, he replied, "Well, that's interesting, I consider my wife my true soul mate, but I consider these guys one with my soul."

Love's Wars

When we love, we become vulnerable to the negative forces within our nature which seem to emerge via our love-lives. The agony and the ecstasy go hand-in-hand because in relationships, our needs and desires must be balanced with those of another. Over the years, very few clients came into my (Paddy's) therapy office asking me to help fix them. It was usually, "I need for you to help me fix—(in the following order): my spouse, my parents, my children, my lover, my boss, my friend, my sibling, my neighbor, or occasionally, my pet. It is always easier to blame someone else for causing us pain than to recognize the deficits in our own emotional and behavioral love repertoires.

Just as we are unable to see the dark side of the moon, so it is with love when the dark side hides within our own psyches. Before we have completed our first year of life, we become afraid that we can't be "good enough" for our caregivers to continue to love us unconditionally. This is when our conflict between love and fear establishes itself deep within our nature.

The conflict becomes obvious within families. Those persons to whom we belonged during our earliest formative years arranged the stage on which our love-lives will be acted out. Due to our genetic make-up, propensities for various behaviors are a biological, hard-wired part of us. There is an identified gene that actually stimulates a greater need for novelty in some persons than in others. The higher novelty seekers are usually the ones searching for blueberries before strawberry season has ended.

There is also a gene that drives some to seek more intensity, more excitement in their lives. These are the daredevils, the ones we tend to think of as courageous, who are usually not given to the most rational behavior. They may be committed to a monogamous life-style, but are usually seeking passion from skydiving or other activities where there is a high mortality rate. They are often not as afraid of dying as they might well be of loving because love requires emotional vulnerability.

Fear

Fear of not receiving enough love and the fear of not being able *to* love are reciprocals. As noted earlier, when an infant has a need, she makes a noise signaling distress. If aid comes to relieve that distress, she feels content, secure, and reconnected. Early attachment love takes root. Research with infants and young children supports the premise that nurturing must be predictably present, so that the infant does not doubt her connection to love. Most care-givers try to meet the needs of their infants, yet some infants have more needs than the care-giver is able to balance appropriately with her (or his) own needs. Even if a mother feels totally exhausted, she might reach deep into her reserve and try one more time to comfort her charge. When the little one responds positively, the mother feels relief and her energy is often somewhat restored. But when the result is not positive and the caregiver feels frustrated and hopeless, she may develop apathy that overshadows her nurturing feelings. This does not mean she feels no more love, but when one's love seems incapable of satisfying our loved one, frustration and anger can make it difficult to behave lovingly. This reciprocal process plays out in almost every form of relationship love. There is a desire to be as content as possible and to want our loved one to feel the same.

However, as difficult as it can be to find happiness within ourselves, it is impossible to find it consistently through another. No one of us can be in total control of another's emotions. All interpersonal relationships pit self interest against the interest of the other to some degree. For example, most parents are rarely consistently perceived by their children as totally adequate, while parents at times will consider their children over-demanding of more attention. In most relationships, this inevitable imbalance leads to some frustration with both members of the relationship. It is not easy to maintain an optimal balance between giving and taking. Some of us will practically twist ourselves into pretzels to try to be whatever our love object wants us to be, but it is difficult to be the right kind of pretzel all the time to meet another's needs fulltime.

Having tried it for several years, I (Paddy) learned it was a love-killer. Better to be myself and be involved with others who appreciate who I am without pretense and performance.

A continued lack of positive interaction in any relationship will eventually cause a break in the love connection. When an infant (or most of us at any age) moves into a space of disconnection from love, the separation is soon filled with de-valuing of self, depression, and fear. Devaluing of self eventually leads to devaluing others and difficulties in developing empathy or compassion are diminished. A poverty of soul— feeling chronically unwanted, disconnected, and helpless to do anything about it—begins.

This state of poverty gives birth to the deepest fear in human nature—the feeling that our existence does not matter to anyone. When we feel assured of our existence through consistent and positive attention of someone who loves us, there is no need to initiate destructive behaviors. But infants who are ignored for lengthy periods of time will often bang their heads against their cribs or bite themselves. Perhaps this self-inflicted pain is perceived as an assurance of life, which is preferable over feelings of non-belonging that are too painful to bear. Despair moves in. In this place of despair, anything can happen—and does. This is why children kill themselves—and others. This is why some turn to violence and other crimes to verify their existence.

Jealousy

This green monster appears when we are afraid of losing love or when we fear there will not be enough love to meet our needs. Jealousy begins early in life when we feel our love-bond with another is threatened. Researchers have proven that babies as young as six months of age often react with rage when their mothers pay attention to another infant, or even a doll, in their presence.[14]

The rage lessens if the mother provides equal attention to her child, but the anger does not completely disappear. When my (Paddy's) second son, Stephen, was born, his brother David was eighteen months of age. A friend had recommended that when I came home from the hospital with the new baby, I bring a doll to David, which might reduce his jealousy. My mother had been keeping David during the few days I had been absent. When his father, baby brother and I arrived, with the doll wrapped as a present for him, he quickly assessed the situation, grabbed his present and clung tightly to his grandmother, totally ignoring his new brother and me. Quickly, transferring the new baby to my mother, I tried to cajole David into allowing me to hug him. He would have none of it. I was to be punished for deserting him as well as for coming home with a new baby.

The next day, seeming to realize that his brother and I were sticking around and having received quite a bit of love and attention from me, David wanted to feed his doll a bottle while I nursed Stephen. Relieved, I felt we were making progress towards diminishing his jealousy. The feeding was followed by bath time. In our kitchen's double sink, I filled both sides with water, and we began the ritual of bathing our babies. After a few moments, I glanced over at David and was horrified to see that he had literally dismembered his doll's arms and legs, which floated around in the sink, while he was trying to pull the doll's head off. Sibling rivalry is witnessed throughout many species. Recall the fratricide of the young owls. Remember Cain and Abel. The sibling bond is a complex one, as sibling interactions result from opposing selective pressures, one favoring competition for parental care, the other selecting for cooperation in fending off unrelated competitors. In the home, where there are no unrelated competitors, competition is the name of the game; on the street, brothers and sisters usually stand together.

When cultural norms and structures enhance the benefits of cooperation and sharing, while reducing the benefits of competition, love can be relatively free of jealousy. In the sixties, my wife and I (Dolf) lived in Kenya and were assisting friends in a surface survey locating and recording prehistoric stone-axes in the Rift Valley, a beautiful scrub desert environment with lions, rhinos and other wildlife. This valley is also the home of the Masai and their herds of cattle. One afternoon, a group of us met an elderly herdsman and three women, who were walking from one settlement to another. We introduced ourselves and, to our delight, received an explanatory introduction in return. Since the man spoke reasonably good English, we surmised that he was a well-to-do cattle herder.

Being properly polite, he introduced us to his female companions.

The first woman he introduced was elderly and simply dressed. She smiled broadly when he referred to her as his first and most important wife, who had borne him many children, lived with him through good and bad times. The second woman, who was carrying a heavy load of household items and other supplies, turned out to be his second wife, whom he called his work wife. He praised her strength and fertility. The third woman was in her early twenties, dressed in a pretty outfit and adorned with traditional jewelry. As she smiled seductively at my male friend and me, her husband introduced her as his play wife. As we went our way and our new casual friends theirs, I mused on the life of these four people, especially the relationship among the wives. They did not exhibit any jealousy or competition. Each woman seemed to know her place in the scheme of things. They more than likely depended on one another and had probably formed a bond of friendship. It was their life strategy and it worked for them.

Crimes of Passion

When the fear of loss of love, which drives jealousy, is fed by suspicion and anger, the psychological stage is set for what our culture calls "crimes of passion." This was the case with Nathan, back in Chapter Three, who shot and killed his wife and her lover. These crimes occur when someone confuses relationship with ownership. The misconception that we own another often turns love into an attempt to control the emotions and behavior of another. Since this is impossible, it is a situation designed for disaster. Because fear, not hate, is the opposite of love, we fight—sometimes to the death—whatever we fear, to destroy it before it destroys us. Too often, rage and violence are ignited by the irrational thought that "if I cannot own and control you, then I'll make sure no one else ever will." The timeless story of King Solomon and the two women who claimed to be the mother of an infant depicts the difference between genuine love and the desire to control another. To determine the real mother, the wise king threatened to divide the infant into halves, so each woman could have a share. The moment one woman burst into tears and said the other woman could have the infant, the dilemma ended. Loving another, above all else, seeks what is in the best interest of the other. Sometimes this means letting go.

Another type of crime of passion is the extreme act of letting go—committing suicide. Suicide, in the name of love, is never committed out of love, but rather out of a profound inability to love oneself—or others. When someone threatens suicide, it is usually a cry for psychological help—a case of being overwhelmed by internal conflicting emotions.

Honesty

In the realm of honesty, several dark shadows engage in fierce battles, for truth is a multifaceted and deeply complex concept. It shifts and changes with new information and in reaction to what might be going on around us. A client recently said to me, "Being honest with myself is often more difficult than being honest with others."

Yet the effort is well worthwhile. As psychologist Roger Gould writes in *Transformations*:

> *The truth, as best as we know it, must be our goal, no matter where it leads us. Every self-deception causes erroneous judgments, and bad decisions follow, with unforeseen consequences to our lives. But more than that, every protective self-deception is a crevice in our psyche with a little demon lurking in it ready to become an episode of unexplained anxiety when life threatens. The self-deceptions which are designed to protect us from pain actually end up delivering more pain. We fortify our deceptions to protect them from the natural corrections of daily life. The larger the area of our mind we find it necessary to defend, the more our thinking processes will suffer. We will not allow our mind to roam freely because new information might contradict our self-deceptions. The larger the self-deceptions, the larger the section of the world we are excluded from.[15]*

Just as we need to be honest with ourselves, we need to do our best to be truthful with others, for without truth, trust is jeopardized and love is under threat. We do not need to spill everything to another, but when we are hiding something that is emotionally starving the relationship and causing us painful emotional conflict, we'd best 'fess up—first to ourselves and then seriously consider what we need to do about our relationship. If the relationship has great deficits, which we've already tried to heal, but it is proving impossible, we're looking at another set of problems. A common defense of lying is to convince ourselves that we are only lying to protect someone from a truth we believe they cannot handle. Often the person we are trying to protect is ourselves from a truth that we would rather deny than struggle through the vicissitudes of its complications. Honesty can require great courage, but its reward is integrity.

Balancing Time and Energy

Maintaining relationship love requires a balance commitment of time and energy so that our individual needs and affiliative needs are each given adequate attention. We could also describe this process as a duel between the selfish and unselfish aspects of ourselves, or the givers and takers that duel within our psyches. The reality is that adult relationships often demand more of our time and energy than we want to give. This dynamic can cause destructive resentment in both partners. It can also bring hidden shadows to the light when partners challenge each other on what they are willing to do for the benefit of a relationship.

We may claim to value a relationship, but if we spend an inordinate amount of time away from the relationship engaged in other activities, our real priorities become obvious. A friend, who was terribly hurt by her husband's long hours on the golf course and limited time at home, essentially presented him with a choice: golf or family. Thinking that she couldn't possibly be serious, he walked out with his golf clubs. But, she was serious. They have been separated for years, although he still has hopes of reuniting—and continuing to play golf. He will have to prove that he has rearranged his priorities before she will even consider living with him again.

Balancing our time between home and the requirements of a professional job is difficult. This is especially true when we want to "get to the top" in a chosen profession. No one tells us that the top can be a lonely place, especially when we're not giving our personal relationships any time and energy. In his book, *Must Success Cost so Much?*, Paul Evans writes that the demands of work involvement often have a negative effect on private life and significant relationships. He suggests that the reason may be not only the pressures of work, but also the reality that many people lack the attitudes and skills needed to make relationships work. It is easier to blame work than to take an honest inventory of other problems and develop skills that will enhance our relationships.[16]

Each of us has the same twenty-four hours in a day. It is what we choose to do with the hours that determines what is really important to us. Having been with several people shortly before their deaths, I have never heard anyone wish for more material wealth. Many have requested another few minutes to tell someone they loved them—or to ask the forgiveness of someone they may have hurt.

Forgiveness

An inability to forgive often exposes our hidden shadows of self-righteousness and blame. These shadows allow us to temporarily feel superior and to forget that we may have made a contribution to the problem at hand, as the late Bob Hoffman, founder of the Hoffman Institute, constantly reminded us in his book with this title: *Everyone is Guilty, but No One is to Blame.*[17]

Love seems to harbor the desire to be completely known and all forgiven, but not always to forgive. Forgiveness means accepting our power to pardon, to extend grace—meaning that we forfeit blaming, judging and feelings of superiority. The game of sinner vs. saint, often played out when there is infidelity in a relationship, is lethal to relationships. Forgiveness is viewed by some as a weakness. In truth, it is a strength. Oskar Schindler said it best in the movie, *Schindler's List*, when he explained to the power-hungry Nazi, Amon, that the emperor who grants pardon to the thief has more real power than the emperor who has the thief executed. Forgiveness is not forgetting. It is a conscious decision not to feed the shadows of blame, self-righteousness, or victimization.

It is a powerful irony that not forgiving always hurts us more than the person we refuse to forgive. Hate tends to harm the hater, filling him with negative feelings that are often expressed through negative behaviors. By contrast, forgiveness opens up the possibility of settling most disputes with respect.

What is Mature Love?

It is not unusual for mature people to question what love really is, for we never seem to outgrow the experience of falling into it. We do not fall in love voluntarily. It happens for myriads of complicated reasons, with the primary one being that our emotional needs are not presently fulfilled. Falling in love adds excitement and passion to our lives, but it only temporarily improves our quality of life. To develop satisfying, mature love for others requires that we learn "to stand in love," accepting others as they are and not wrap them in a fantasy of who we want them to be. In short, we must realize rather than fantasize them. It is said that men marry women thinking that they will never change—and we always do. Women, on the other hand, marry men thinking that our love will change them. It rarely works. There does seem to be some truth in this adage, however, the deeper truth is that we all change with age and experience. When we love another, we appreciate the uniqueness of his or her aging process.

We value their needs and desires as much as we value our own. We are willing to compromise, to sacrifice when their needs are greater than ours and to listen to what they express with an open mind and an open heart. Our focus is on whether or not the other feels loved by us rather than whether we feel loved by them. We tend to feel loved when we behave lovingly. We need to be able to trust the people we love not to hurt us intentionally, but we must understand that loving requires being vulnerable and honest. This means that we will hurt at times, but love gives us the strength and courage to deal with whatever we must—and to forgive. Love wants the other to be as passionately alive as possible and to have what is necessary for their happiness—even though we may not be included. Mutual love involves sharing our hearts, minds and souls in a communion respected by both partners. Love and lust are not one and the same, yet, if lust and sensuality are appropriate aspects of the relationship, we tend to feel more passionately in love. There can also be deep and lasting love without sexual involvement.

Mature love is not restricted to lifetime commitments. True friendship, within or outside of sanctioned commitments, may be the best form of love. It is created by two people who are committed to sharing who they are and what they have with each other for any period of time. Love is never possession of another, but a continuing choice to relate to another as honestly and completely as possible. Above all else, mature love is compassionate love. In his acceptance speech of the Nobel Peace Award in 1984, the Dalai Lama stated that peace demands we live in the world with compassion, which he defined as ultimate love. Compassion allows us to remain open-minded and open-hearted.

When Will Durant, author of the eleven-volume *The Story of Civilization*, was asked what he considered the most valuable information he had learned from his studies and his life at the age of ninety-two, replied: "If only we could learn to love one another, there need be few other lessons."[18]

References:

Q. Kopp, Sheldon. *Kopp's Eschatological Laundry List*. In *Hidden Meanings*.

1. Tobin, James. *Ernie Pyle's War: America's Eyewitness to World War II*. New York: Free Press, 1997.

2. Smith, Adam. *The Theory of Moral Sentiments*. In Oxford University Magazine, 1759

3. Lewis, C.S. *A Grief Observed*. New York: Harper Collins, 1961.

4. Keen, Sam. Lecture on *The Passionate Stages of Life*, Washington, D.C., 1999.

5. Harlowe, Harry F. "Basic Social Capacities of Primates," in *Human Biology*, 1959, Vol. 31.

6. Bowlby, John. *Attachment and Loss*. New York: Basic Books, 1969.

7. Gray, John. *Men are from Mars and Women from Venus*. New York: Harper-Collins, 1992.

8. Psych. Today. June, 2000. (validated via several medical encyclopedias)

9. Tennov, Dorothy. *Love and Limerance.* New York: Scarborough House, 1999.

10. Moore, Thomas. *Soul Mates.* New York: Harper-Collins, 1994.

11. Goodall, Jane, with Berman, Phillip. *Reason for Hope.* New York: Warner Books, 1999.

12. Sevile, George. In Bartlett, John. *Familiar Quotations.*

13. Lawrence, T. E. *Lawrence of Arabia.* In *Seven Pillars of Wisdom.* Library Edition. Castle Hill Press, 1922.

14. N.Y. Times Magazine, April, 2002.

15. Gould, Roger. *Transformations.* New York: Simon & Schuster, 1978.

16. Evans, Paul and Bartolome, Fernando. *Must Success Cost so Much?* London: Grant & McIntyre, 1980.

17. Hoffman, Robert. *Everyone is Guilty, But No One is to Blame.* Oakland, California: Recycling Books, 1988.

18. Durant, William. In Wall, Ronald. *Sermons for the Holidays.* Grand Rapids, Michigan: Baker Book House, 1989.

Promises and Problems of Belonging

No man is an island, entire of itself.
~ John Donne, 1624

While love may hold the ultimate answer to our survival as a species, it is not an easy answer. To master our ability to love and to have our needs for love fulfilled may be our greatest challenge. Love makes promises that are often broken, and even the most loving relationships are at times beset with frustrations, misunderstandings and conflicts. Yet, when all is said and done, none of us would have survived to date unless someone, or a group of others, had loved us enough to protect and nurture us through our beginnings. Because we are social beings, we continue to need love and to seek it through relationships to individuals and to groups—beginning with our families and extending to large organizations, to nations and belief systems. We need to belong to someone and be connected to a tribe, so to speak, in order to develop a personal identity and to fully experience our aliveness. Many prisoners of war have reported that being placed in solitary confinement for extended periods of time was more difficult to bear than torture. People do need people.

In the previous chapter, we introduced the concept of belonging love from the standpoint of belonging to other individuals. Now, we turn our focus to the drive to belong to a group of others. This drive is already apparent in simple grazing animals—such as buffalo or muskoxen, which have a desire to stay within a herd. These low-intelligence animals have evolved a kind of love for belonging to a group, even though they show little or no attachment to a particular herd, nor to any particular individual in the herd beyond the cow-calf—bulls ignore all calves. Since in evolution new traits usually originate via natural selection modifying existing traits, it is likely that the love of belonging to a group arose like pair-bonding love, from selection acting on components of material love and kin relationship love.

This evolved step towards love for unrelated individuals of one's species is a very important one, because it is the first step towards bonding with other individuals for reasons other than sex (procreation) or cooperative behavior based on kin-selection. The biological advantages that must have driven the evolution of a desire to belong to a herd, tribe, pod, flock or school include safety, better access to resources, warmth and sharing work. These benefits can activate strong levels of group dependency. Pigs, rats, geese, primates and humans are among those who will pine away until they die when isolated for extended time periods. We will see how this belonging love gave rise to more complicated interactions in species that have developed higher levels of culture as part of their group dynamic.

In humankind, belonging love is a hugely complex enterprise with major implications for our individual lives, as well as for our social, cultural and political structures. This type of love inherently comes with promises and expectations for various levels of commitment. It almost always gives rise to some disappointments and problems. To understand

176

the implications of belonging to a group, including the promises and problems that tend to ensue, we will again visit some of our non-human relatives to learn how individual animals negotiate their sense of being a member of a social group.

Life in the Treetops

In February 2006, I (Dolf) spent a couple of weeks in the rainforest of northern Borneo, studying forest management and its effect on wildlife in the region. This offered me an opportunity to observe some exceptionally beautiful animals in their native habitat. On one such occasion, I was with a few friends on a small boat on the Mananggol, a narrow river surrounded by giant trees. These trees spread their branches above and across the deep dark water, creating bridges for arboreal animals, such as the leaf monkeys, orangutans and snakes, to cross the river. The river's edge was richly blended into the forest with a variety of low vegetation, while rafts of floating water hyacinths with their jacaranda blue flowers added a touch of brilliance to the greenery. Every now and then, we had to skirt past a tree that had fallen into the river. This maneuver was not to be executed carelessly, as the frequently still living branches of a partially submerged tree were a favorite place for green pit-vipers to snooze away the daylight hours, and these venomous serpents do not like being rudely awakened. Just past a gentle curve in the river, we came upon a troop of proboscis monkeys, the largest of the leaf monkeys. They were busy doing what they do for most of their lives, which is foraging on the leaves of the trees they inhabit.

We were excited to find these rare and spectacularly grotesque-looking monkeys. Most were mature females, juveniles and young adults, all overseen by an alpha male. The females and young adults had exceptionally long, retroussé noses, which made them look quite comical and more human than most monkeys. The alpha male particularly captured our attention with his appearance and attitude, projecting a persona of complacent arrogance. His pink, swollen, drooping nose was enormous, his tail was long and snowy white, and his belly bulged above a large, permanently erect, fire engine-red penis and a coal-black, rotund, well-filled scrotum. To fill out the picture, imagine him lounging comfortably on a couple of branches above the river, glancing here and there, keeping an eye on his troop, appearing even more colorful then he really was, as the late afternoon sunlight gave his rust-colored pelage a golden hue.

Quietly observing this troop of monkeys made it clear to me that each individual had its own personality and identity, which was not only an expression of its innate characteristics, but also of the prescribed role each was scripted to act out within the troop. The monkeys seemed to be consciously aware that expressing their position in the troop was an important part of their identity. The alpha male was obviously in control. He kept an eye on the females and young adult males, who kept a goodly distance from the females—and an even longer distance from the alpha male.

In an environment of abundant food, such young males are an asset to the group, as they take on the role of scouts around the periphery of the troop, while it moves through the forest. As long as they keep their distance from the females and perform their scouting role, these males are "promised" safety as members of the troop. The females kept busy taking care of the juveniles, who played close by the females, or clung tightly to those who gave them nourishment and love. Within each category, there were distinct differences in individual behaviors.

There was also evidence of dominance hierarchies within the subgroups, as females and lower ranking males had minor altercations over choice feeding spots, or over who could approach which juvenile. Our main conclusion was that these animals, like most advanced social species, had a strong sense of belonging to this group as an essential aspect of their individual identities. Simultaneously, they demonstrated a sense of belonging to sub-groups, from the simple mother-infant pair to the juvenile male community on the periphery of the troop.

Life Beneath the Trees: the Basics

Our personalities and identities are also formed through a sense of belonging to a number of overlapping and hierarchical groupings, each of which has an effect on our behavior. One of the most influential groups to which we belong is our family of origin. Family members, especially parents, tend to keep their unspoken promises to meet our needs, which establishes a dependency on them and a desire to stay connected to them.

Each of us is affected by our position in our families, be it a family of origin or a family of choice. Within the structure of the family group, we learn to obey, to manipulate, to understand boundaries, to defend ourselves and often to challenge authority. At a young age, we begin to identify with the values of our parents and to understand that being a member of our particular family means we are expected to behave in a certain way and above all, to pledge our allegiance to "the family."

We learn which behaviors bring rewards and which bring punishment. We are taught to defend and protect members of our family and are usually rewarded for doing so. The ways we view ourselves are deeply embedded in our psyches through the politics and dynamics of our family, and our continuing need to belong to a group is reinforced through being in a family, regardless of how dysfunctional the family may be.

There has never been a perfect family due to the complexities of conflicting interests among family members and the conflicting emotions within each of us. Families seem to develop unique ways of elaborating on these conflicts. One of the most poignant scenes of the film, *The Lion in Winter*, illustrates the power of convoluted family politics. The three sons of Eleanor of Aquitaine and Henry II are arguing with their father over whom he should choose to be the next king of England. Richard, the eldest, a homosexual and his mother's choice for the throne, is holding a sword to his father's throat, while John, the youngest and his father's choice, is having a temper tantrum. The middle son, who knows he is not the choice of either parent, is loudly complaining to his mother that her deviousness has divided their family. Henry decides to disown all three sons and have his marriage annulled on the grounds that Eleanor has probably slept with his father. She waltzes from the room with the ultimate understatement: "Well, I guess every family has its ups and downs."[1]

She has spoken the truth; however, there are extremes of dysfunction at both ends of this continuum. When children are born into severely dysfunctional families and no one responds to their needs, they will attach to anything available, since the need to belong is critical to continued development. According to a report of a Texas social service agency in the late 1990s, a carnival-working mother kept her baby in a large cage in the back of a truck, shared with several dogs. When he was able to eat solid food, he was fed scraps in a dog dish slid under the door of the cage. His mother tried to put him up for sale when he was eighteen months old, which caught the attention of local police, who intervened. After the toddler was placed in a foster home with caring parents, he cried endlessly until he was put in a room with a trained dog, who became his constant companion. After establishing a belonging relationship with the dog, he was gradually able to accept his foster parents and identify himself as a member of the family group—which included his beloved dog.[2]

To progress as a viable member of society, every child needs to belong to someone who fulfills the promise of caring for him. Our early childhood experiences form a template for our life patterns of interacting with others and of our behavior in groups. It's as if our membership in our first group establishes a kind of shadow box in which our perceived realities, memories, illusions and fantasies of our families continue to play out in all subsequent relationships, including the groups with which we will become involved.

Life Beneath the Trees: the Reality

As we mature into adulthood and find our own place in society, we have more choices about our associations. A man may be a committed member of a family, an extended family, a neighborhood or community, a political party, a men's social group, a pipe band, a professional organization, an ethnic minority and a nation. He will have a sense of belonging to each of these groups and share a special connection with other group members. He will also experience various levels of belonging love and commitment to each group and will contribute accordingly to each group's welfare. To some degree, he may be controlled by his association with one or more of his groups. And, like Mozart's Tamino, from Chapter Three, some people not only promise to sacrifice their lives for the good of a group, but actually do so.

In December 2006, Ross McGinnis, a nineteen-year old Private First Class in the United States Army, gave his life to save four other members of his platoon. As the group rode in a military vehicle in Baghdad, McGinnis was perched in the gunner's hatch when a grenade sailed past him into the truck. He shouted a warning to his comrades in the truck before throwing himself over the grenade, which exploded, killing him instantly.[3] This kind of commitment to a group may sound extreme, but it is the level of commitment required by war. War holds the possible promise of glory jointly with the more probable problems of violence, killing and death. World War II Japanese kamikaze pilots and twenty-first century suicide bombers are other examples of this extreme sacrifice. Wars are about one group killing another, and all members of both groups must be prepared to kill and to die.

From childhood on, we develop our personalities and identities in a complex sequence of groupings to which we belong. Some of these groupings, such as race and gender, are biologically determined and are more or less permanent. But they invariably will carry a cultural load. Being a woman, for example, is not only a matter of being physiologically a female, but also makes you a member of a society that imposes limitations, opportunities and responsibilities established by the cultural norms of your society on women. Membership in other groups, such as religious congregations or social organizations, may be imposed upon us in childhood, but as we become more independent, we are able to choose more freely whether to make a commitment to such groups. Whether our affiliations are biologically determined, chosen, or forced through coercion—for example, being drafted into the military—we base our identities and much of our behavior on our membership in groups and on our perceived status within those groups.[4]

Like the leaf monkeys, our love for belonging is so strong that we are willing to make sacrifices to protect the groups with which we identify, for they are essential to our social identities. We will pay membership dues, argue in the broader political arena for our groups to be officially recognized, participate in organizing events and willingly give our time and energy to help other group members. In times of crisis, people who feel that a group to which they are strongly committed might be threatened will go to great lengths to preserve the group's status because it reflects their personal status. Some religious institutions have stretched their moral ethics to protect members of their clergy who have sexually assaulted minors in their congregations because it reflects badly on the status of the entire clergy.

This willingness to fight for our strongest love connections can be essential to our emotional and physical survival, but it can also become destructive. The conflicts between different groups within a society and the conflicts among individual members of subgroups as they jockey for status within their groups form the basis of human politics. Understanding politics will expand our understanding of why we wage war. In order to develop a more objective view of our political instincts, a look at the political behavior of our closest four-legged relatives might prove worthwhile.

Chimpanzee Politics

Some of the most vivid descriptions of social interactions among chimpanzees are reported by anthropologist Frans de Waal in *Chimpanzee Politics: Power and Sex among Apes*. In this book, de Waal describes the individual personalities of members of a large captive group of chimpanzees and their interactions over several years. Whereas this community was isolated in a zoo in Europe, making contact with other communities impossible, a striking finding was that these apes behaved in much the same way as described for communities in the wild. The essence of this behavior was a self-imposed limit on internal aggression and violence. Whatever the internal relationships demanded from individuals, the overriding reality was that the community was a self-contained, sovereign entity, totally dependant on its defensive ability for survival. De Waal found that even in this isolated zoo community, males fighting for social status in the power hierarchy made a great deal of noise, beat one another up and even bit each other's extremities, but only rarely did they use their formidable canines to inflict serious injury.[5]

In chimpanzees, especially in males, natural selection has been unremittingly severe in selecting for a supreme sense of belonging to the community. While human athletes, aspiring to stardom, can risk losing a few games due to hyper-competitive behavior within their team, chimpanzees in the wild cannot afford to weaken their community to the point of losing border skirmishes with neighboring communities. Nonetheless, this dynamic does not stop the desire for power and status within the group. For animals as intelligent and politically savvy as chimpanzees, it means that they, like us, form less stable sub-community coalitions that are of great importance in determining the power structure of the larger community. These lower levels of belonging are also important determinants of each animal's identity.

The role of coalitions is clearly described by de Waal in his reports of the takeover of the alpha position by one male from another. Reading de Waal's rendition of this power struggle is like following a soap opera on television. Episode by episode, the drama unfolds, with coalitions formed and broken, allegiances changed, and intimidation and manipulation freely employed. Sooner or later, overt violence usually begins. As the power structure starts to shift, individuals change their roles in the social structure and adjust their personalities accordingly.

The two main characters in de Waal's chimpanzee drama are Yeroun and Luit, mature males who were accepted as numbers one and two respectively in the overall dominance hierarchy. Yeroun, the older of the two, was the alpha male, and Luit was his buddy. Luit and all other community members showed their submissive status by greeting Yeroun with the standard greeting with which lower-ranking apes greet higher-ranking ones whenever they come into close contact. As Yeroun became older and less agile, Luit became less submissive to him. One day, Luit initiated a confrontation by blatantly mating with a female in heat right next to Yeroun. This opened a 72-day period of increasingly aggressive interactions between the two leading males. At first, the females sided with Yeroun, but after being bullied by a third male, Nikkie, who had taken on the role of Luit's henchman, they started to accept Luit as their leader. At the end of this period of instability, Yeroun gave up and acknowledged Luit as the alpha male.

What makes this story more than a one-on-one fight for supremacy is that Luit, despite his greater strength, could not have succeeded without the support of a henchman and a coalition of females. Nikkie did not join directly in any of the fights between Luit and Yeroun; his strategy was more subtle. He attacked some of the females, which normally would have resulted in disciplinary action by the alpha male, but Yeroun was too occupied by his conflict with Luit to restore order. The result was that the females gradually lost confidence in Yeroun as an effective leader. Was this a conscious strategy planned by Nikkie, and the resulting shift in the females' support his intention? It is hard to see it differently. One by one, the females started to abandon Yeroun, and eventually the old alpha male reluctantly acknowledged his new role as number two by greeting Luit as the number one male.

The Evolution of Our Belonging Identity

Perhaps the most important aspect of the above chimpanzee saga is that the chimpanzees seem to be aware that their survival depends upon both competition and cooperation. Although the competition for the alpha male position was harsh, once the new leader was acknowledged, peace returned to the troop and all members assumed their new roles. This cohesion would be necessary for their survival if they were living in their natural habitat and had to protect themselves against neighboring aggressors. The fact that they lived in captivity did not change their instincts to increase their fitness through competition within the group and maintain their group's survival quality through cooperation. They dealt well with "the hockey player's dilemma," as described in Chapter Four.

Since proboscis monkeys and chimpanzees in the wild have highly group-dependent individual identities, we should be able to pin point the ecological and social factors that form the natural selective forces driving their strong sense of group belonging. Assuming that our ancestors over many millions of years have lived under similar ecological and social conditions, we can study their behaviors to gain a better understanding of our own belonging emotions and identities. In any socially structured group, the balance between cooperation and competition provides the basis of the opportunity to gain both individual status within the group and increased shared fitness due to group cohesion. More specifically, the group provides more efficient resource management, better protection, a secure breeding environment and the opportunity for each individual to play a role best suited to his genetic propensities. A group is most effective when there is intrinsic diversity, which allows an optimal degree of division of labor. For non-human primates, genetically determined propensities for hunting, fighting, leadership, care giving, foraging and tool making would be the prime traits for frequency dependent natural selection to act upon.

For humans, there is an infinite list of traits, skills and talents that demand intra-population genetic diversity. Too little diversity could lead to an unbalanced group structure, but too much diversity could also be disruptive. Too much competition would lead to intra-group violence, while too much cooperation and tolerance could lead to the survival and propagation of undesirable traits, such as complacency and passivity, which could destroy ambition and passion. In every generation, natural selection acting on the members of the group will re-adjust these two main aspects of group structure and cohesion.

In non-human primates, sub-groups, such as small political alliances, female groups or juvenile male groups within troops, add complexity to the social system, while being an important part of the larger community. Such communities are themselves sub-groups of an entire population, and unfortunately, the strongly antagonistic relationships between communities are also a major contributor to the evolution of belonging identity. The typical chimpanzee population structure depends on individual males having developed a sufficiently strong sense of belonging that will motivate them to embark on the risky business of raiding their neighboring communities with the intent to kill when they feel threatened. The fear of losing resources or power feeds perceived threats. Between each pair of groups or subgroups lies an identity gap; the wider the gap, the more distrust and hatred can build up between members of the two groups, and the more likely it is that lethal violence will erupt when conditions bring the groups into a highly competitive situation. What makes inter-group competitive situations especially volatile and dangerous is the escalating fear of differences, as explained in Chapter Five.

This is also the case for us humans, due to an innate fear of the differences which we have not reconciled with the behaviors and customs more familiar to us. Most humans are not likely to settle differences between competing badminton clubs or political parties with overt violence, but when perceived survival or the power to control the larger community are at stake, the fear that we will lose control, power or resources seems to fuel the most destructive aspect of our competitive nature. Among Muslims, this concept of an escalating and ever widening identity gap has been played out for over 1,300 years between the Sunnis and Shiites. They appear to be more alike than different in the eyes of the rest of the world since they live in the same country and worship the same god, but the fear of "the other" has escalated to the point that they hate and kill members of the other group.

The history of civilization records many examples of homogeneous societies splitting into factions that adhere to different religious interpretations and practices. In England, during the Tudor period, the members of early Protestant groups were burned at the stake until Henry the Eighth broke with Rome, after which Catholics faced the same fate. This lack of tolerance for differences escalates fear and hatred, often leading to wars—in which different proselytizing religious groups try to gain power over others. This is becoming widespread between Christians and Muslims over the world as we write.

Family Politics

Because our family units are among the strongest groupings to which we feel a sense of belonging, and there is usually less of an identity gap among members of the same family, we tend to try harder to keep our propensity for violence under control. This does not mean that there is no competition for status within the family, but that for the purpose of "looking good" and feeling accepted by the larger community, we tend to try to hide our ugly warts from "the public." What makes the family such a powerful unit is the combination of the genetic relationships, the early-life social cohesion most people experience and the ways in which we carry our nepotistic support for our family members into higher levels of social organization.

Often, we make promises to help our family members and count on their promises to help us. Each individual uses her place in the groups she belongs to/or influences to further her status both within the groups and in society at large. She will also further the status of her relatives or fellow group members and use their influence to her own advantage. This survival-driven strategy of furthering one's own and one's relatives' interests inevitably implies stepping on, or at least blocking, other peoples' interests. Those, whose interests are threatened, will more often than not fight back, just as the chimpanzees do.

186

Our conflicts between cooperation and competition may be every bit as dominant as they seem to be in these ancestors, but we have developed more subtle ways of managing them, based upon what worked for us within our families. The father is usually assumed to be the head of the family—the alpha male, but he can be manipulated by a wife who can either support him or undermine him. We cannot afford to forget that we all need to have our existence acknowledged and to feel some degree of personal efficacy. We each tend to ally with those family members who most meet these personal needs, and we form subgroups accordingly. At times, adolescents, seeking more freedom, will join together to "make war" against the power of the parents. When the parents disagree, each parent is apt to select the child, or children, most prone to agree with them, to form an alliance against the other parent. Thus, we are perpetually torn into opposite directions between cooperation with family members to advance our joint interests and competition with other members to gain personal power, love or extra favors.

We have evolved masterly sets of social and psychological skills within our families to maintain a balance between our competitive and cooperative natures that stay with us throughout lives. When family dynamics go awry and overt violence results, there is frequently a high price to be paid, the effects of which reach far beyond our immediate families.

Gangs, Blood Feuds and Vendettas

When early-life love, care and social cohesion are not adequately provided, young people may grow up without a sense of belonging to a family and instead focus their thwarted desire for belonging love on clans, gangs, clubs or religious/ethnic groups, which demand high levels of attachment but provide low levels of love and security. Alternatively, tight families attempting to create a rewarding place for themselves in alien circumstances and/or disorganized societies will frequently escalate conflicts with outsiders. In each case, the ethnic, religious or family "gangs" fail to fit into a smooth hierarchy of levels of belonging. When insecure people feel threatened, they are likely to create deep identity gaps between themselves and members of neighboring groups, which to them represent a perceived threat. Inevitably, they in turn become a

187

threat to others, as is the case between the Sunnis and Shiites—and many other similar situations the world over.

In the mid-1800s, when central Canada was gradually being populated, central government was weak and disorganized. Many of the area's inhabitants were recent immigrants of various European countries, some of whom had recently experienced revolutions, foreign occupations and collapse of the rule of law. Under such circumstances, it is not surprising that local sub-communities in central Canada formed gangs for the purpose of exploiting others, or became vigilantes in defense of such gangs. From 1847 until 1880, the town of Lucan and surrounding areas were terrorized by a gang of Irish immigrants known as "the Black Donnellys." Theft, robbery, murder, arson and intimidation kept the region in constant fear. Ineffective marshals and the unwillingness of witnesses to come forward allowed the Donnellys a free hand to carry on as they wished for years.

Finally, in 1880, a vigilante group retaliated. On a cold February night, a group of drunken men descended upon the Donnelly homestead and butchered five of the gang members, while the remaining ones got the message and disappeared. Interestingly, whereas on previous occasions when the law had tried to deal with the gang, fear of retribution had silenced witnesses called against the Donnellys, the vigilantes also got off because no one broke the silence. No reliable witnesses to the murders could be found, so the jury sided with the murderers. It is worthwhile to speculate about the identities of these Donnellys.

They were a large family consisting of a father, a mother, a bunch of sons and hangers-on. Their strongest sense of belonging was to the family, with a secondary attachment most likely to their Irish background. As recent immigrants, their sense of identification with the new country and its citizens was minimal. Following the general rule that once a member of your group is in trouble with people beyond the group, your loyalty to the group member overrides any responsibility to other groups, the entire Donnelly clan stuck together in defense and considered all others as enemies—and therefore, as exploitable.[6]

Such gangs as the Donnellys are known from other parts of the world— especially in young societies that are disrupted or stressed. The current twenty-first century version of the alienated sub-national gang is the youthful street gang, the motorcycle gang or the recent immigrant-based mafia. A fast-growing and far more dangerous group is the alienated, underground, religion-based terrorist cell. These groups are usually led by a fiery, charismatic leader, whose followers are often sons of immigrants who, despite a fair level of education, have failed to integrate comfortably into mainstream society. They typically see their outsider status as the result of religious or racial discrimination, especially in the areas of career development, employment opportunities and acceptance in desired social circles. It has come as a shock to liberal democracies that some well-educated, young middle-class men feel sufficiently alienated to withdraw from mainstream-society, join groups of similarly alienated men and build up sufficiently high levels of disgust and hatred, to motivate them to execute acts of random terror against their fellow citizens.

Warlord-led paramilitary groups and clan militias are another type of 21st century sub-national group, which is prevalent in Somalia, Afghanistan, and other developing countries that lack strong central governments. These groups often have histories of feuding over regional power that has lasted for centuries. Historically, the feud between the troublesome MacDonalds and the shrewd and ruthless Campbells is perhaps the best-known blood feud known in the English- speaking world. Both clans date from the thirteenth century and remained mortal enemies for at least four hundred years. For most of that time, the Scottish clans operated more like Yanomamö villages than modern nations, in that they recognized other clans as belonging to their nation, but their sense of commitment was primarily focused on their own clan.

After the massacre of Glencoe in 1692, any hope of peace between these two clans was lost. That winter, on a cold, snowy day, Robert Campbell ordered a troop of his soldiers to "...fall upon the M'Donalds of Glencoe and put all to the sword...". Thirty-eight men, women and children were murdered, with many more freezing or starving in the hills where they fled. Until this day, and probably far into the future, Campbells and MacDonalds may try to make light of the feud, but feelings of resentment likely remain. We should also acknowledge that the Campbells and the MacDonalds are not the only Scottish clans that have carried on blood feuds. Many smaller clans have been virtually wiped out in repeated clashes.[7]

Another word for blood feud, which has crept into the English language from the Italian, is *vendetta* (from the Latin *vindicta* for vengeance). For longer than most other European countries, Italy remained a patchwork of city-states and small kingdoms, with little or no central government power. As these independent political entities grew in size, they depended more and more on the surrounding countryside for a reliable food supply and tax base, and as a defensive perimeter. However, increasing the control by city-states over the surrounding region brought the *grandi* (dominant city families) into direct conflict with the *contadini* (the landed gentry), resulting in long-lasting family feuds that became known as vendettas.[8] One of the best known of these was the feud between the Montecchi and Capello families of Verona, which became the inspiration for Shakespeare's Montague and Capulet families immortalized in *Romeo and Juliet*.

In a somewhat more civilized day and age, these conflicts that arise from destructive family politics, both intra and inter, are regulated and resolved via family and state courts, which can limit the scope of the violence. However, the real seat of these conflicts still lies deep within our individual psyches. We want to belong to a group of others. We seek the safety and resources offered by becoming members of larger organizations, but we are reluctant to compromise our personal needs and desires, when they are in conflict with those of the group.

The Dark Side of Belonging

As described earlier, our closest relatives, the chimpanzees have a strong sense of belonging to their communities, yet feel absolutely no sense of connection to chimpanzees of neighboring communities. The term "xenophobia" for their intense antagonism is not appropriate here because, for chimpanzees, it is not a phobia—it is the norm. It is a strongly established trait of the species, driven by their belonging love for their community, selected because those who were willing to kill and risk being killed had, on average, a higher fitness than those who wavered in this duty to their fellow group members. This is probably so because participating in raids has a political pay-off in the individual's status in the community, where status implies fitness. Understanding the relatively simple political system of the chimpanzee is critical to the understanding of our own political system and our predicament of charting a course for a peaceful future. The chimpanzee's intra-community politics of jockeying for power, while keeping the lid on lethal violence, counter-balanced by what we might call a foreign policy of genocidal violence, may not be a perfect analogue for our species. But there are some intriguing—and troubling—similarities.

What we know of primitive hunter-gathering societies suggests that when their ecology was like that of the chimpanzees, they had similar political structures. When their ecology was less attached to defendable resources, simple avoidance of neighbors was more likely than overt warfare. We do not know in detail how hunter-gatherer tribes made the transition to the gradual growth of civilization, or how they adapted their social behavior to an agricultural economy. When we again consider the Yanomamö, the primitive horticultural society discussed in Chapter Four, we find a situation similar to the chimpanzees, but with a more complex and sophisticated approach to internal politics and foreign policy.

Yanomamö societies are only loosely hierarchical, some almost totally egalitarian, but invariably still sufficiently hierarchically structured to be polygynous. The killing rate within the village community is generally low, consisting mostly of impromptu murders from jealousy or anger and occasional planned executions of disruptive members of the clan by majority agreement. Yet, violence and war between communities, and especially between tribes, is often very severe.[9] The Yanomamö have a complex social structure with strong sub-village alliances or groupings based on kinship and inter-family marriage. As the village population increases in good economic times, the feeling of belonging to sub-village groupings can increase while the sense of belonging to the village as whole may decay. Eventually, such villages will split into two separate ones—each large enough to provide the members with at least a modicum of protection from external enemies. This splitting of large communities has also been observed in chimpanzees.

Anthropologist Napoleon Chagnon in *Yanomamö* describes in detail the intense psychological pressure experienced by the males of a Yanomamö village when a decision was made to launch a raid to

191

avenge their neighbor's past aggression. For two days before the raid, the village performed traditional ceremonies including symbolic cannibalism, obviously aimed at reinforcing the men's sense of belonging, their anger at the enemy, their manly pride and their commitment to participate. Nonetheless, many inexperienced fighters showed fear mixed with bravado, and once the raid began, several showed cowardice, but masked it by pretending to be ill or injured. The ceremonies also had the effect of making the raid inevitable, by committing the participants before any danger could change their minds. Many women were equally ambivalent, in both supporting the raid and weeping out of fear for their sons and husbands.

Chagnon's description of the prelude to the raid shows how strong the sense of belonging to the village is, but also how an established system of authority can manipulate and coerce the people into dangerous acts of aggression. No one is born with a sense of belonging to a certain village, nation or any other group. We are born with a genetically determined propensity to want to belong to the group in which we grow up and to be swept up by the social outrage against an identified enemy, when group members feel threatened. This trait has undoubtedly saved many villages, clans and tribes from defeat by their enemies, but it also forces individuals to pay the price for being the front-line defenders of their communities. Worse, it opens us up to manipulation by unscrupulous leaders, mass hysteria—or both.

Enforced Nationhood

When we make the leap from rainforest horticultural societies to agricultural and even industrial states, the outbreaks of large-scale violence among geographic or ethnic subgroups become more frequent and severe. Human history of the past couple of millennia has seen the formation of many states that bring together various smaller nations, tribes, imported slaves, conquered neighbors or otherwise semi-sovereign communities under a central government. This usually eventuates into long periods of violence, suppression, enforced integration, coercive education and the eventual formation of a national culture, to which most citizens feel an overriding sense of belonging.

Most European nation-states were formed in this manner, and many third world countries are currently moving in the same direction. This gradual shift from belonging love for the smaller tribal cultures to a national one has been a very violent process, which, even in twenty-first century Europe, is far from complete; witness the independence movements in Corsica, Scotland and among the Basque people of northern Spain, as well as failed federal states such as Yugoslavia and The Soviet Union. Meanwhile, war-like aggression between sub-national social groups, with members of each group having a much stronger sense of belonging to their clan, tribe or region than to the nation-state, is still a normal state of affairs in most developing countries.

But even established countries, such as the United Kingdom, Canada, France and Spain, have ethnic minorities clamoring for more rights and autonomy, if not for outright independence. Examples of current violent suppression are not hard to find: e.g., Russian Chechens, Turkish Kurds and Palestinian Palestinians. Not only has lethal violence reigned frequently within incompletely formed states as governments refused to recognize and have attempted to destroy the national ambitions of minority groups, but violence between states, in the form of war, has also been a normal phenomenon. The centralized governments promised peace and stability, but usually tried to deliver such promised goods by having the majority tribe exploit the minorities and by expansion into neighboring territories.

Multiculturalism and the Open Society

During the twentieth century, many secular liberal democracies have committed themselves to high levels of individual freedom within multicultural societies. This development was based on the need to integrate inflows of displaced persons and immigrants after World War II, as well as the need to accept equality rights for established, ethnic and religious minorities. In post war Europe, the inflow of immigrants from behind the Iron Curtain and from ex-colonies created major social problems for the established nation-states, especially the United Kingdom, France, Germany and the Netherlands.

At the same time, ethnic groups, such as the Friesians in the

Netherlands and the Welsh in Britain, demanded the recognition of their languages. Many such problems were addressed and alleviated with increased legal recognition and protection of linguistic, ethnic, racial and religious rights, as well as a clearer definition of individual human rights. However, major stresses remain, as members of minority groups often fail to identify with mainstream society and still experience discrimination. Even Canada's integrated multiculturalism and America's melting pot concepts are far from achieving their desired unifying effects.

In most European countries, a malaise of pessimism has set in as the established secular majorities perceive an eagerness by some immigrants to take advantage of the economic opportunities of their new homes, while refusing to accept the secular community's norms. Following riots in three industrialized cities in Britain in 2001, a Government investigation found that growing social and religious segregation, based on separate educational, volunteer, religious and cultural institutions, largely triggered the violence.

In the Netherlands, the wake-up call came on November 2, 2004, when the filmmaker Theo van Gogh was murdered on an Amsterdam street by a member of a radical Muslim organization. It was not a random, isolated murder. Van Gogh had recently released a film about the oppression of women in Muslim society. The murderer had skewered onto Van Gogh's body a four-page diatribe, which began with an ominous threat: "Beware Madame, as a soldier of evil, you are doing the work of the enemies of Islam..." and continued to promise that when they, the killers, caught up with her, she would be praying for her death.

The police and the Dutch people knew exactly to whom this note was addressed. The real target of the killers was Ayaan Hirsi Ali, a member of the Dutch parliament, who was born in Somalia. She had fled to the Netherlands and had become politically active, working for the rights of women and publicly criticizing the treatment of Muslim women. Her latest "crime" was to have written the script for van Gogh's film. When the police raided the suspected murderers' organization, they uncovered a plot against Ms. Hirsi Ali that included a plan to kidnap, torture and kill her. Ms. Hirsi Ali was provided with twenty-four-hour bodyguard protection and was driven in a bulletproof Mercedes to and from Parliament, where she insisted on continuing to work for women's rights. She no longer resides in the Netherlands.

It is difficult for members of a free and more secular democracy to understand the predicament of those in more traditional societies. Whether severely repressed women in traditional Islamic societies, Indian untouchables, or simply a Catholic woman who wants to become a priest, breaking out of limited roles in rigid societies more often than not causes rejection and even prosecution. When faced with this alienation and punishment, they often give up their dreams to remain "stuck" in the horrors of a place that secures their belonging.

The Dutch people and their government still believe in a multicultural society, but refuse to grant sub-national organizations the right to violate the human rights of their members within the larger secular society. Without condoning the murder of van Gogh or the oppression of women, we need to make an effort to understand the mental state and resulting criminal acts of some conservative Muslim men. They feel that the foundation of their faith and culture are being destroyed by evil Westerners and want the balance between individual rights and religious group rights to be skewed much further towards group rights than the Netherlands, or any other western secular country, is willing to accept. They have demonstrated that they are willing to kill for what they consider their right to protect their religious group.

In 2004, the Islamic Institute of Civil Justice in the Canadian province of Ontario requested from the Government to have the right to use legally binding religion-based arbitration to settle family disputes dealing with such issues as divorce, custody, alimony and inheritance. The request almost slipped by unnoticed, but later in the year, when an official report recommended the Government grant this privilege with strict conditions and safeguards, a storm of protest exploded. Major objections were voiced, mainly from other Canadian based Islamic organizations that feared that such arbitrations would lead to violation, mostly of women's rights. In the forefront of this campaign was Ms. Homa Arjomand, who is a Toronto based refugee from Iran.

She organized an international campaign, which resulted in major public demonstrations, not only in Ontario, but also throughout Canada and in many European countries. By February 2006, the Government had read the public mood clearly, and instead of granting the privilege, they refused the request and simultaneously took a similar, previously granted privilege away from Jewish and Catholic groups.

In principle, this conflict is very similar to the one described above for the Netherlands. It deals with the same conflict between the traditional rights of religious groups to restrict individual rights and impose restrictions on the freedom of their members and the more recent laws of the secular, liberal democracies of the West, which guarantee individual freedom to all citizens, regardless of which religious group claims them as their members. As could be predicted, this conflict is far from settled while new challenges and counter moves are already in the daily news. Recently Ms. Arjomand said:

> We must not rest, we must continue to fight for human rights, freedom of speech and the benefits of a secular society, not just here, but also overseas. People in other countries who live under burdens of Sharia and Political Islam, countries such as India, Iran, Iraq and Nigeria; these people are counting on us.[10]

Unfortunately, we can expect more anger and violence from some religious groups. Even in educated, civilized societies, the desired concept of implementing a system of mutual multicultural respect in a free, secular society runs into serious problems.

In his now famous book, *The Open Society and Its Enemies*, the philosopher Karl Popper foresaw such problems when he argued that a free secular society needs to find an optimal level of freedom. He expressed his belief that not enough freedom will lead to failure, while too much freedom will also lead to serious problems, for the very reasons we are now experiencing. In liberal democracies with multicultural populations, Popper's paradox of democracy surfaces in an unexpected and critical new way.

How can we guarantee freedom to the individual, while at the same time granting freedom to organized minority groups that deny individual freedom to their members? Not only do we have this philosophical paradox, but also a political situation which maintains deep identity gaps between society as a whole and self-ghettoized minorities. Individuals must have the right to join any group and commit their belonging love to that group. If a similar level of belonging love is not extended to a wider, inclusive society, and if the group demands rights beyond what is considered reasonable by the larger society, we face the dark side of love, which leads inevitably to conflict and violence.[11]

How we can establish this balance of freedom and reasonable limits on violence on a global scale remains a hope, a dream and a desire until some seven billion people of different cultures in a world of growing resource shortages are willing to allow their love of belonging to the human species to override their love for claiming a larger than fair piece of a shriveling pie for the local tribe or nation with which they primarily identify.

Global Prospects: Promises and Problems

At the international level, two political trends have gained growing adherence since the Second World War. In Europe, we have seen an ongoing process of forming negotiated federations of open, multicultural states with reduced sovereignty, in which rights of intra-state, as well as inter-state, minorities are guaranteed. So far, this process must be considered a major success, as it now seems inconceivable that war would ever be waged again between, say, France and Germany. Yet, too much optimism is still inadvisable, as it is yet to be seen whether most Europeans will be able to transfer their primary sense of belonging to the European Union from their individual nationalities when economic or environmental stresses increase.

The other major development is a growing and genuine desire for lasting worldwide peace. Over the past half century, the utopian concept of the Global Village has been applied to various goals, from vague calls to extend understanding and compassion to all peoples, to specific proposals for forms of global government. Some real progress has been made in this direction, in the form of United Nations resolutions and treaties dealing with human rights, refugees and the treatment of prisoners.

The United Nations, numerous countries and some international

NGOs monitor the status of human rights and occasionally succeed in shaming violators into ceasing some of the most grotesque violations. On the ground, however, everything from outright mass genocide and so-called "pre-emptive" wars, to the torture of individual political prisoners and state-sanctioned assassinations, are still almost as prevalent in this 21st century as in the past.

We began this chapter with the thought that "Love may hold the ultimate answer to our survival as a species…" We realize that the call to develop compassionate love, defined as the ability to remain open-hearted and open-minded to all others, may sound far too idealistic to be acknowledged as a realistic goal. The sacrifices required by each of us to initiate the necessary changes might go beyond what we are willing to give—or give up. And yet, if we want to maintain a sustainable world for our species, we have reached a point at which we must commit our minds, hearts and energies towards creating a safer and more peaceful world—in which compassionate love has a chance to thrive.

References:

Q. Donne, John. in Bartlett, John, *Familiar Quotations, 13th Edition.* New York: Little, Brown, and Company, 1955.

1. Harvey, Anthony, *The Lion in Winter.* Study Guide, BookRags. The Gale Group, Inc.. Farmington, MI, 2000.

2. Texas Social Services Report, 1997.

3. Associated Press Release, Nov., 2006.

4. Appiah, K.A., *The Ethics of Identity.* Princeton University Press, 2005.

5. de Waal, Frans, *Chimpanzee Politics—Power and Sex among Apes.* Johns Hopkins University Press, Baltimore and London, 2000.

6. Kelley, Thomas P., *The Black Donnelly's.* Firefly Books, Buffalo, 1993.

7. Thomson, Oliver. *The Great Feud—the Campbell's and MacDonald's.* Sutton Publishing, 2001.

8. Muir, Edward. *Mad Blood Stirring: Vendetta and Factions in Friuli during the Renaissance.* John Hopkins University Press, 1993.

9. Chagnon, Napoleon A., *Yanomamö: The Fierce People.* Harcourt Brace Publishers, 1997.

10. Kingston, Ontario Daily News and CBC reports.

11. Popper, Sir Karl, *The Open Society and Its Enemies.* Fifth Edition. Routledge & Kegan, 1966.

CHAPTER

War and Peace Revisited

It is easier to make war than to make peace.
~ George Clemenceau, French Prime Minister, 1919

Joyeux Noel, a French film produced in 2005, portrays an unusual Christmas Eve celebrated by World War I soldiers in northern France. On that evening in 1914, some soldiers of the British Expeditionary Forces were patrolling along the parapets of the trenches facing the no man's land separating them from the Germans. As they kept watch, the soldiers made efforts to celebrate Christmas by singing carols and opening packages from home. Makeshift Christmas trees decorated the parapets on both sides of the front. As the British sang the last notes of "God Rest You Merry Gentlemen," the Germans began "O Tannen-baum." Within the first few notes, the voices of all the troops blended together in "Stille Nacht," during which a lone German soldier began walking with a lighted candle across the dangerous divide toward the other side. He was soon met by a British soldier, who invited him and his troops over to exchange food, drinks, and cigarettes.

As the camaraderie spread down the front, an impromptu soccer game was enjoyed between teams of German and Scottish soldiers, who shared a barrel of beer at the conclusion of the match. The joint celebrations lasted until Christmas Day, when word reached the commanding officers of the British, Scottish and French troops. Outraged, these officers ordered an immediate end to the festivities and threatened to court-martial anyone who did not comply.

Over the next few days, the war-gods roared and the carnage resumed, as many of the fun-loving youths of Europe were killed by those with whom they had recently enjoyed celebrating Christmas. The following year, when word spread that several British soldiers were planning another Christmas Eve truce, they were immediately court-martialed. By 1916, during the nine-month Battle of Verdun, the French suffered 315,000 casualties and the Germans 280,000. When the armistice was signed to end World War I on November 11, 1918, the majority of the soldiers who had participated in the 1914 Christmas truce were dead.[1]

In writing about this wondrous Christmas Eve celebration, Wilfred Owen, the revered poet of World War I, proclaimed that this brief interlude proved that brotherhood and compassion might have a chance to overcome "the pity of war." The price would be giving up the myth that war is about honor and glory and seeing war as it is—the product of greed for more resources and for domination over others.[2]

Following the end of this Great War, heralded as "the war to end all wars," due to the use of weapons of depersonalized mass destruction, resulting in a staggering loss of young men's lives, the way the world viewed war was changed forever. Most of our lives have been touched by some aspect of war, which has shaped our divergent opinions and beliefs about it. But in the words of a survivor of Iwo Jima, during World War II: *Every jackass thinks he knows what war is, but no one knows until he's been in one.*[3]

Voices throughout History

This soldier's words hold a ring of truth, but nevertheless, every citizen of any country at war risks suffering the consequences of war—including the realities and prevailing fears of losing loved ones, having to sacrifice resources and of losing sovereignty. It is time to raise our personal awareness of the price we all pay for war so that when there are future calls to arms, we will not, as citizens, blindly support a leader who advocates waging war, and thus be responsible for the heinous crimes sanctioned through war. The diversity of attitudes toward war and peace is vast; however, reviewing a few of the extremes propounded should give us pause to consider our deepest beliefs about war and peace. Evaluate your personal reaction as you read.

The nature of war consisteth not in actual fighting, but in the known disposition thereto.
~ Thomas Hobbes, English philosopher, 1651.[4]

For what can war, but endless war still breed?
~ John Milton, English poet, 1648.[4]

Are we then to give up the sacred graves of our ancestors to be ploughed for corn? Dakotas, I am for war.
~ Sioux Chief Red Cloud, 1866.[4]

There never was a good war or a bad peace.
~ Benjamin Franklin, U.S. statesman and inventor, 1783.[4]

This war is an inconceivable madness.
~ Maude Gonne, Irish nationalist, 1914.[4]

War is the most exciting and dramatic thing in life.
~ Moshe Dayan, Israeli general, 1972.[4]

Everlasting peace is a dream, and not even a pleasant one; war is a necessary part of God's arrangement for the world.
~ Count Helmuth von Moltke, Prussian general, 1893.[4]

We are now suffering the evils of a long peace. Luxury, more deadly than war, broods over the city, and avenges a conquered world.
~ Juvenal, Roman satirist, *ca* 115.[4]

Peace hath higher tests of manhood than battle ever knew.
~ John Greenleaf Whittier, Quaker poet, 1853.[4]

I am sick and tired of war. Its glory is all moonshine. It is only those who have neither fired a shot nor heard the shrieks and groans of the wounded who cry aloud for blood, more vengeance, more desolation. War is hell.
~ William Sherman, U. S. Federal general. 1870.[4]

War is a force that gives us meaning, but it is not a uniform experience or event. It gives meaning to sterile lives; it also promotes killers and racists. It creates powerful self-deceptions, but when we grasp war's reality, a universe collapses.
~ Chris Hedges, American war correspondent, 2002.[5]

In the above quotes, we hear desperation in the acceptance that war seems to be an unavoidable festering wound on humanity, the opinion that it is an important part of our nature, the determination to use it to defend one's interests and the admission that it can stir our passions, especially when we become apathetic or restless. Lasting peace also seems to have been experienced as a mixed blessing—and for some, even as a curse. It is easy to assume that most members of the human race would prefer to live in a peaceful world, but our history and several authors, including Tolstoy in his classic, *War and Peace*, reject that assumption. It is time to consider these incongruous attitudes and begin to understand how both war and peace can be aspects of human nature, and why we have not been able to put an end to unjust wars—if there is such a thing as a just war.

The 17th century English philosopher Thomas Hobbes can be excused for failing to understand that human nature is not a fixed state, but rather a highly variable, constantly evolving set of traits with a long evolutionary history. The expression of human nature is also dependent upon social and economic conditions. Possibly due to the extremely violent and insecure conditions in Great Britain during his life, which included the civil war and Scottish and Irish rebellions, Hobbes took a dim view of human nature. He saw mankind as primarily brutish and in need of strict training to become civilized. Milton and Clemenceau expressed the more generally accepted point of view that war is somehow inevitable. When Sigmund Freud was asked if he believed war was inevitable, he agreed with Milton and Clemenceau, based upon his conviction that our nature is basically selfish. Franklin and Gonne saw no benefits in war whatsoever, while Dayan and Juvenal defend war as a means of eliciting passion and arousing citizens from lethargy. General Sherman and the Iwo Jima survivor share the view no one can possibly understand war's deplorable reality without actually having fought in one.

Why are there such widely divergent beliefs about war and peace from highly educated people throughout recorded history? Are we really so different one from the other, or is what we proclaim to believe based upon our personal experience and we are prone to change our beliefs contingent upon varying experiences? Could it be that our nature is to judge most situations as relative to their opposites? For example, do we need to experience war to appreciate peace? Are we prone to create conflicts for stimulation after extended peace, as suggested by Juvenal? Is it possible that we have not yet explored peaceful alternatives to war deeply enough to be able to resolve problems without resorting to the ways of our past?

In previous chapters we have discussed the ubiquity of war throughout human development and its persisting prevalence in our present world, despite major international efforts to reduce its frequency—and eventually to abolish it altogether. The most promising developments are the relatively stable, peaceful blocks of nations in Europe. However, in many parts of the world, things continue to drift from one war to the next. The major questions we must seek to answer are:

- When economic and social conditions deteriorate, will mankind by nature inevitably resort to war as the ultimate solution to its problems?

- Will it be possible to convince governments and people who seem eager to wage war to seriously consider peaceful alternatives to resolve future conflicts?

- What price are we, as individuals, willing to pay for peace? Or, in the final analysis, which wolf will dominate?

Human Nature under Scrutiny

In order to better understand the intricate complexity of our relationship to war and peace, we will discuss a rare experiment that sheds some light on an aspect of human nature that could control the balance between peaceful coexistence and warfare. A team of psychologists, sociologists and social anthropologists interested in exploring the complicated dynamics of inter-group conflict and cooperation experimented with two groups of pre-pubescent boys under isolated campground conditions. Their objective was to show how social conditions can quickly establish group-specific behavioral norms of either overtly antagonistic attitudes or of tolerance and friendly cooperation with other similar groups. The 1954 study became known as the Robbers Cave Experiment, due to its location adjacent to Robbers Cave State Park in Oklahoma.[6]

The experimental design was simple and effective. Two groups of ten members, all eleven-year-old boys, were selected so that they were as homogeneous as possible in terms of personal developmental, sociocultural background and family stability. Although the boys lived in the same city, care was taken to assure that they were not previously acquainted with anyone else in the groups. They were taken to a large campground where each group was exposed to normal camp conditions in separate parts of the campground, each with its own swimming hole and other facilities. So, for a few days, each group remained unaware of the other's existence.

During the first week of the three-week experiment, each group was given a series of challenges, as well as ample free time, while the professional staff surreptitiously observed and recorded behaviors. Not surprisingly, the groups soon developed a social structure, with a dominance hierarchy, and with each boy beginning to play his own role in that structure. The interactions among the boys were typical of any semiegalitarian political system with some jockeying for status; lower-status boys paid tribute to the leader, who in turn either criticized or praised them for their actions and ideas. There was manipulation, consensus formation, conflict and conflict resolution. During the week, group-specific norms became evident and identifying symbols, flags and group-specific jargon, were adopted.

Toward the end of the first week, the two groups were moved within closer proximity so that each became aware of the other. Their immediate reaction was antagonistic, assuming that the other group was somehow a threat to their current status on the campground. Significantly, at this stage, each group chose a name, one the Eagles and the other the Rattlers. When the staff verified the presence of the other group, the boys in each group referred to the others as intruders and labeled them with derogatory terms, such as "dirty bums" and "sissies." An atmosphere of "us versus them" became prevalent, and both groups expressed a desire to challenge the "intruders" to a baseball game, which each group assumed they would easily win.

During the second week, the staff organized a series of competitive situations between the groups, including baseball, tug-of-war and tent pitching. As the days progressed and each team won and lost some events, the degree of antagonism increased rapidly. Unbeknown to the boys, the staff had made efforts to assure a tie of wins and losses between the teams. On both sides there were now expressions of hatred and contempt, with each team accusing the other of cheating. Before the end of the week, overt violence exploded. Scuffles, for which some boys had armed themselves with socks containing stones, broke out, forcing the staff to intervene before these weapons could be employed. Cabins were raided, their contents scattered, and stolen flags were burned. During the raids, some boys were gung-ho to attack the enemy, while others fearfully remained some distance behind. By the end of the second week, both groups expressed a desire to cut off all relations with the other group.

During this week of competitions, within-group cohesiveness had become much tighter than during the first week. Each group became possessive about their own territory, and the boys bragged about their group's stellar performances (whatever the reality), while devaluing their opponent's performance. There were also changes in the dominance hierarchy in response to how individuals dealt with the aggressive, competitive events. The report of this second week is, in principle, not that different from Chagnon's report of the Yanomamö's raid on a neighboring village or relationships among neighboring villages in New Guinea, especially when we take into account that the staff prevented the use of dangerous weapons.

Due to the escalating negativity between the Eagles and the Rattlers, the staff faced quite a challenge in making the transition to the third week, during which the experiment was to test the boys' response to finding themselves in a camp-wide vexing situation that could only be alleviated by having the two groups cooperate. The experimental hypothesis was that cooperation, which could benefit both groups, would diminish their antagonism and even initiate some positive interactions. The staff shut off the water supply and told the boys that the pipe from the reservoir to the camp was probably damaged. They were then told to organize a way to locate and fix the problem. This, and similar complex challenges, gradually brought the two groups together, as they had to confer to lay plans and then cooperate in implementing them.

By the end of this third week, the two groups still maintained separate names, leaders and social structures, but they had begun to cooperate effectively. No more between-group antagonism was expressed, and past events, such as the raids on the cabins, were now discussed in mixed groups with shared humor and accolades for individual acts of courage or initiative. Relations between members of the Eagles and the Rattlers varied from cordial to friendly. A cooperative state of peace had been created— a dramatic transformation.[6]

The Conundrum of Human Nature

Certainly, what was possible on a campground with two small groups of eleven-year old boys with an overseeing staff would be more difficult to achieve on a larger scale with heterogeneous groups competing for life-sustaining and limiting resources. We might well ask what the outcome of the third week would have been had there been only enough water for one group. Would the boys have negotiated sharing the water and all gone to bed somewhat thirsty? What would have been the situation at the end of week two had the stone-containing socks been used and resulted in serious injuries? Perhaps, without adult staff supervision and direction toward pre-determined goals, these boys would have become as destructive and savage as the lost gang of boys in William Golding's *The Lord of the Flies*. In this classic novel, Golding takes the pessimistic view that the beastly side of our nature would dominate if it were not held firmly in check by society.[7]

While we cannot afford to over generalize the results of the Robbers Cave experiment, it does indicate something important about human nature. The first week demonstrated the desire and need of these boys to belong to a group. When thrown together without recourse to the security of their original groups of family and friends, they promptly formed a group *de novo*. It probably helped that they already had several things in common due to their similar social backgrounds, but there was also enough individual diversity to provide each group with leaders and followers. Each boy brought a well-developed identity to the experiment and quickly adapted that identity to new circumstances. When conflicts arose within a group, they were settled amicably or with nonviolent social pressure. Each group developed a surprising degree of balance between individual freedom and gently coerced conformity. Might we conclude that this study represents a conformation of human nature at its best? Perhaps, but we must remember that during the first week, there was plenty of good food, comfortable shelter and protection provided by a mature camp staff to allow the boys to feel safe and free of the need to compete strongly within the group.

With the discovery of a second group, another aspect of human nature revealed itself. With strangers nearby and the camp staff intentionally doing nothing about these "intruders," the underlying insecurity of each boy, having only recently been separated from their established support system, was powerfully triggered. Within the group, differing opinions and the expression of hitherto unexpressed traits led to a reshuffling of the dominance hierarchy and a tightening of expectations to conform to group norms. Feelings toward the other group were strongly antipathetic, and a desire, based on overconfidence, to beat "the enemy" and attain supremacy became the driving force of each group.

As the second week progressed and competitive situations were presented, resulting in fairly high levels of frustration, the antipathy grew to a level that could have resulted in violence, had the staff not intervened. Worth considering also is that if the staff had not arranged the competitive events in such a way that the final scores for the groups were tied, would both groups have wanted to end the competitions?

This stage of the experiment demonstrated in a microcosm what has happened countless times on a larger scale when groups—from villages as primitive as the Yanomamö to countries as large as the USA— sense an external threat. The vilification of a perceived enemy and the desire to unite as a group to subdue or destroy that enemy usually result in a willingness to forego some individual freedom. The idea that there is safety in banding together seems to be a deeply ingrained aspect of most social species.

While kayaking on a lake in Ontario recently, I (Paddy) saw two seagulls, perched on a rock, protruding above the lake's surface, emitting loud screeches. Within moments, these two were joined by four others, and as if on a given signal, they took flight simultaneously. Knowing that seagulls are not always socially cooperative, I was fascinated, so stopped paddling to watch. Their screeches became louder; they looked and sounded like a bomber squadron, as they chased a bald eagle, one of the most feared predators of still flightless gull chicks, away from the lake and out across the treetops. The gulls qualify as a flocking species in that they cooperate collectively to protect their breeding colonies, whereas more socially sophisticated species, including ourselves, also protect feeding territories—not only from marauding animals, but also from neighboring communities. We have evolved under social and ecological conditions which gave a fitness advantage to those who chose to unite and fight rather than give up access to resources and face the downward spiral of starvation, pestilence and persecution.

The most applicable finding of the Robbers Cave experiment to our exploration of whether or not we humans will ever be able to eliminate war was the transformation that took place during the third week. But unfortunately, this result is the least reliable. Although the staff was meticulous in not telling the boys how to think or behave toward the members of the other group, they did create situations that brought the boys together in non-competitive arrangements. This made it relatively easy to overcome animosity once the sources of frustration were removed.

Worthy of note also is that whereas the social, educational, ethnic and religious similarity of the boys in both groups made the antagonism between the groups in the second week more significant, that same similarity made the ability to make peace much more feasible. Following serious conflict, peace may not be so easily achieved with people from entirely different cultures. The experiment provides evidence that the boys, and by extrapolation, people in general, might make peace when they no longer feel threatened and do not have to make major sacrifices or adjustments to their social order.

This conclusion is reinforced by what happened spontaneously along the French-German front on that memorable Christmas Eve—a willingness to cooperate and make peace with the enemy. What makes this WW I event especially significant is that these soldiers were literally in mortal battle prior to their shared Christmas celebration. To walk out of their protective trenches to meet in the open of no man's land indicates that their desire to create an interval of fellowship and peace was strong enough to overcome their fear of death. Also significant is that whereas the boys in Oklahoma were aware of their leaders' encouragement, the soldiers were willing to risk the ire of their superiors to temporarily reduce the madness of the war.

What about Peace?

In Chapter Five, we discussed the prevalence of war in human societies throughout, at least, the past thirty-five thousand years, from small band hunter-gatherers to more complex agricultural communities to more recent industrial societies. But, what about peace?

As noted in the Introduction, Tolstoy claimed that love and war offer the only true passions in life and that peace is a boring state of affairs. Has the idea that peace can also be a passionate experience been given short shrift? Concluding that a propensity to fight wars is an aspect of human nature does not rule out that a propensity for making and trying to keep peace is also an important aspect of our nature. Long periods of peace have prevailed between wars in many parts of the world.

In recent literature, conflicting opinions have been voiced on the relative frequency and duration of war and peace over the ages. Some books argue vociferously that contrary to the accepted opinion that primitive hunter-gather societies were generally peaceful, they fought wars frequently, while other books argue the exact opposite.[8] After analyzing these arguments, it becomes clear that the differences of opinion are mostly due to different definitions of war used by the writers, as well as their limited knowledge of hunter-gatherer ecology, demography and evolutionary biology.

Our working definition of war for the purpose of this chapter is: a conflict with lethal violence between two sovereign communities, which directly or indirectly involves a large portion of the members of both communities, either as combatants and/or victims. This includes civil wars, raiding and border skirmishes, but not individual acts of violence related to sub-community level conflicts. There is also a gray zone when group violence erupts within larger communities among ethnic, religious, criminal and regional populations when they reject the established laws and norms accepted by the larger community. We will simply define peace, again for the purpose of this chapter, as the absence of war. When reviewing the literature using the above definition of war, it is obvious that both war and peace take their turns in most societies.

War is generally feared, yet frequently becomes considered inevitable. This is the case in a defensive war to protect the survival or sovereignty of a nation, and in an offensive war to limit another group's power or to gain access to valuable resources. Peace negotiations and formalized conflict resolutions between communities to maintain peace are also common features of most human societies.

Lawrence Keeley stresses the peaceful attitudes encountered in small tribes found in isolated regions of difficult terrain. More than likely, many of these tribes have moved into such areas as an escape from violent interactions with more powerful neighbors. After being defeated by a strong enemy, fleeing can often be the best survival strategy. An aggressive neighbor loses a strategic advantage relative to the defenders in difficult terrain like mountains, swamp forests, and deserts and will gain few advantages by invading what is likely to be an unproductive area.

The tribe which has fled into the inhospitable landscape will need to adjust to what could be a dangerously structured environment with new ecological challenges, but they gain an opportunity for a peaceful survival. As a result of this situation, many such small, isolated tribes suffer sufficiently high mortality, which will prevent them from over-populating their resource base. In South America, some of the Yano-mamö villages in the mountainous southern range of their tribe seem to be this type of isolated, peaceful community.[9]

Another example is a cluster of small tribes, communally known as the Upper Xingu from Brazil. These people cannot really be consid-ered one tribe because they represent four different language groups and live in ten politically sovereign villages. Yet, the villages live in relative harmony with each other and have developed an interdependent trade system, contingent upon non-violent interactions for its success. They are occasionally raided by non-Xingu neighbors, forcing them to band together in defense.[10]

An interesting history of long-term peace, revealed in archeological research by Steven LeBlanc, involves the native people of the North Amer-ican Southwest from 500 AD until Western contact. His work indicates that from 500 AD until 900 AD the population size of the region was relatively stable, due to a dry climate, which resulted in periods of famine. There was also evidence of frequent warfare, possibly associated with these famines. Then, around 900 AD, the climate became considerably wetter, opening vast areas of previously arid wasteland to agriculture. Over the next two hundred and fifty years, the population increased fivefold, villages and cities spread over the region, and for the entire period, there is no evidence of war. Towns had no defensive walls, and buildings were not built so that they could be defended when attacked.

These two and a half centuries of prosperity, population growth and peace were the result of new climatic conditions creating an ecology of plenty. Then, over the following two centuries, the climate became dry and cold. The land could no longer sustain the now five times more dense population, which crashed, as starvation became common and vicious wars raged. The deteriorating agricultural productivity caused the inhabitants to decide that war was their only way to survive, and those who did survive were the ones most capable on the battlefields. This bit of history teaches us that when ecological conditions allow for peace, we humans are more apt to cooperate and avoid war, but we are capable of fighting wars when our survival is threatened.[11]

Japan, after a long period of civil strife among regional feudal shoguns (warlords) in the sixteenth and early seventeenth centuries, also entered a peaceful period of over two hundred years. By 1637, Tokugawa Ieyasu, the strongest of the shoguns, had defeated all others and united the country, which he and his successors ruled as a centralized feudal state, with the shogun in charge and the emperor as a figurehead head of state. Internal peace was maintained within a rigidly imposed social structure, enforced by the hierarchical shogunate, in principle, not that different from what now would be called a military dictatorship.

Relations with neighboring countries and seafaring European traders were kept at a minimum, as the government's strategy was one of nearly total isolationism, and their potential enemies did not possess the regional military might to force Japan into changing this policy. For the first century of the Tokugawa shogunate, the economy and Japanese culture flourished. However, a merchant middleclass rose and began to demand the right to trade internationally. As their political power grew, the central government weakened. From the middle of the eighteenth century, the country went into decline.

When several European countries and China became aggressively involved in demanding trading rights, and Russia occupied parts of Sakhalin Island, the U.S. government decided to take the initiative. President Fillmore ordered Secretary of State Daniel Webster to dispatch warships under the command of Matthew Perry. The official reason for this act of aggression was to "civilize" Japan after the news that Japan was persecuting its Christian minority. The true motivation for this act, that the United States was searching for an acquisition in the Pacific, allowing for increased steamship trade, was not revealed until Perry was well on his way. On July 8, 1853, four U.S. Navy ships with sixty-one cannon entered Tokyo Bay. In this land where steamships were unknown, the fear that these black-cloud-belching monsters, resembling evil, fiery dragons, would destroy their way of life became a reality. Within the next fifty years, Japan's government changed drastically, embracing industrialization and international trade at an incredible pace.

By 1904, Japan had defeated China and Russia with their own armed steamships. For over two centuries the Japanese had lived in peace due to their military power and their isolationism, but after 1853, the reign of peace in Japan ended.[12]

War and Peace: Two Sides of One Coin

It is fair, during this period of our evolutionary development, to suggest that most humans prefer to avoid war, but that sovereign communities, when afraid of losing resources or freedoms, find themselves willing to fight. As intimated by Clemenceau and Whittier, it is easier to trust what has proven to be a sometimes-successful strategy (war) than to run the risk of trying potentially peaceful alternatives, which could demand a higher test of manhood. Currently living populations of our species are the descendents of those who have not only fought wars, but have been victorious. Over all the millennia, we have been selected to be warriors. And even during peaceful periods, we could not afford to neglect our defenses because war could be forced upon us.

Evidence of extensive earthworks, moats, stone ramparts and other defensive structures discovered in prehistoric sites are often taken as an indication of war, but some of these defensive structures could have been an attempt to keep peace. We know that after the War of 1812, during which American troops invaded Canada, the British built Fort Henry at the place where the Saint Lawrence River flows out of Lake Ontario as a deterrent for any further American incursions; it was never attacked. Similarly, in the twentieth century, the ultimate deterrent was the nuclear ability of the USA and the Soviet Union for mutual annihilation, which continues to hold us at bay. In all probability, defensive structures of the past and present have partially succeeded to avert some wars. Through the ages, we are aware that efforts have been made to maintain peace, but peace has a weakness, in that it does not select against those who chose war.

Those who fought with superior skills and possessed the more deadly weapons would win and leave their progeny to carry on. Thus, we have evolved under a set of environmental and political conditions and have developed responses to those conditions which have led to a higher biological fitness than other, less pugnacious, responses. When threatened, or tempted by an opportunity to gain more resources by threatening others, we usually respond by becoming bellicose. This is the way of evolution.

To change the ways we have evolved is impossible, but it is critical to be aware that evolution is geared primarily to survival through competition and not to produce unerringly ethical or moral human beings. Because we have also evolved a conscience as an important aspect of our higher consciousness, we often pay a high psychological price for war. We both have elderly friends who served in WWII and never talked about their war experiences until decades later in their lives, when each began to share feelings of ambivalence and deep conflict over their time spent on the battlefields.

One friend explained that because he had served in the artillery and therefore never came face to face with the enemy, he was able to convince himself that he possibly never killed anyone. Now, facing the end of his life, he realizes that he has been deluding himself for years, and his feelings of guilt are in conflict with his having carried out his duty to his country. Another, whose hair turned from brown to white soon after the invasion of Normandy, confessed a week before he died that what happened there was not deserved by anyone on either side of "this atrocious war." He admitted that he had prayed for months that God would forgive him for his participation in the slaughter on that beach, and that for the remainder of his life, he felt guilty for being alive. "During that particular battle," he said, "it was as if the soldiers were no longer humans, but killing machines gone wild."

These are not unusual stories. War and peace are not only two sides of the coin of survival, but are two primary aspects of human nature, which take turns expressing themselves contingent upon circumstances at a specific time. The fact that we now live in a world with enough weapons of mass destruction to extinct ourselves several times over, the psychological conflicts may well force us to seriously examine less violent ways of resolving the conflicts threatening our survival. In these next sections, we will take a look at a human society as far removed as possible from our own for the purpose of studying their ways of conflict resolution, as applicable to war and peace.

War and Peace among the Murngin

Those anthropologists who argue that a propensity for waging war is not a basic component of human nature do not deny that mankind has been fighting wars ever since the coming of agriculture and its concomitant social and political structures. Rather, they focus their argument on the pre-agricultural era, when all people were hunter-gatherers. Recently, archeologists have shown convincing evidence of much pre-agricultural warfare, as well as evidence of periods of peace. To better understand the role of war and peace in these early societies, it is worthwhile to take a look at some of the anthropological research of the early twentieth century, which focused on the behavior of Australian aboriginal hunter-gatherers.

The aboriginal communities of northern Australia's Arnhem Land are typical of many hunter-gatherer tribes. Before the establishment of Western government control, they were reported to live mostly at peace with their neighbors. Nonetheless, suspicion and real or imagined grievances simmered close to the surface, and occasionally erupted into acts of overt warfare, in the forms of frontier battles, raids and ambushes. When the anthropologist W. Lloyd Warner studied the Murngin tribe in the early twentieth century, the tribe was split into twenty politically independent clans of some forty to fifty individuals. Whereas the entire tribe did not function as a political entity, each clan inhabited and defended its territory. Clans were led by headmen who had only limited power over what were fairly egalitarian communities. Each territory was centered on one or more waterholes, spiritually and ecologically of the highest importance.

A key influence on the tribe's social dynamics was a complex polygynous mating system in which all male members were fully committed by tradition to attract wives only from clans other than their own, which effectively externalized intra-clan violence. What little intra-clan violence Warner encountered consisted either of isolated murders or of consensus-based executions of habitual troublemakers, those who showed no regard for the clan's established norms and rules. When men from two clans competed for the same women from neighboring clans, or when conflict arose over which clan owned a certain waterhole, inter-clan violence could become intense.

During the time that Warner studied the Murngin, it was reported to him that two neighboring clans, the Warumeri and the Wangurri, had recently fought two border wars referred to as *gaingar*, which resulted in twenty-nine deaths. The most remarkable aspect of this report was that the two clans had assumed that this extremely lethal *gaingar* would perma-nently settle their differences, and therefore end all future wars. Unfortu-nately, it didn't work out that way. Warner calculated the long-term, war-related male mortality rate for the Murngin at twenty-five to thirty percent, remarkably similar to estimates for many tribal societies in other parts of the world. Yet, the clans were frequently involved in discussions about how to settle inter-clan grievances and made use of ceremonial methods of conflict resolution, including the *makarata*, a form of non-lethal combat between selected members of each community.

The initiative for a *makarata* was taken by an aggrieved clan who wanted to end a feud with a neighboring clan instead of seeking a revenge killing. They would send a message to their enemy proposing a *makara-ta* at a certain time and place. The opposing clan usually accepted, but both sides would be nervous and suspicious up to, and during, the entire ceremony. On the day of the contest, the men of each clan, their bodies heavily painted with white clay and fully armed with spears, faced one another on the selected site. They sang and acted out in dance the spirits of their waterholes or performed the dance of one of their sacred animal totems, often, the crocodile.

The *makarata* was usually performed on a level piece of ground adjacent to bush, so that the men could flee and take cover if things got out of hand. After the ceremonial displays, a complex sequence of spear throwing was executed, while the aggrieved party yelled insults to the opponents held responsible for the most recent killing. The ceremony ended with the intentional thrusting of a spear into the thigh of the accused killer(s). If the process went according to plan, and no one got seriously injured, or killed an opponent or a bystander, the *makarata* was considered successful, and peace between the two tribes could prevail for several years.

On the other hand, if something went awry, the *makarata* ceremony could incite violence rather than prevent it. Warner's description of the Arnhem Land tribes illustrates that these tribes make serious efforts to live in peace. Men from different clans may fight over women, but such conflicts are often settled through *makarata* ceremonial combat. Disagreements over the ownership of waterholes, however, could lead to serious war, as neither of the antagonists is willing to give up this essential life-sustaining resource without a fight.

One factor which describes the intricately complex Murngin inter-clan relations is that their strict adherence to cross-clan mating is both the cause of inter-clan violence and of the potential for reducing violence and negotiating for peace. This is because cross-clan mating results in members of any one clan having close relatives in other clans. Murngin society is a patrilineal kinship as well as a clan divided society. Political power lies in the clan, but personal identity is also affected by one's kinship. At times of conflict, the clan could decide to execute a raid, but several clan members would not want to harm those opponents who happened to be their relatives.[13]

Why the Peaceful Walbiri Wage War

It is unfortunate that scholarly interest in hunter-gatherers began after these people had already been influenced by more structured societies, resulting in few reliably documented cases of hunter-gatherer societies living strict traditional lives. We have already seen that the Yanomamö of South America were horticulturalists, depending mostly on non-indigenous food-plants and were killing one another with steel weapons. Similarly, the Murngin have been trading with Malay traders for centuries who provided them with metal objects. By the time Warner studied them, they were already fully aware of the presence of the Australian government and powerful, unsavory western intruders, such as missionaries and gold prospectors, but they remained exclusively hunter-gatherers.

The Walbiri of the Tanami desert region in central Australia were another native tribe who retained an exclusively hunter-gatherer way of life during this same period. They survived in this extremely arid region as hunter-gatherers until well into the 1950s, when anthropologist M.J. Meggitt lived among them to study and describe their way of life. The overriding factors in their relationship with nature are water and drought. Whereas the Murngin suffered most during the excessive rains of the monsoon climate, the Walbiri had to eke out a yearlong survival from a climate with very short rainy seasons separated by long, extremely hot droughts.

This relationship between people and a sparse, unforgiving environment is reflected in a very low population density (estimated by Meggitt at only fourteen hundred individuals on approximately fifty thousand square miles) and a nomadic occupation of various areas of their territory throughout the year. They are only loosely separated into four subunits, which tend to stay in their own sections of the tribal range. But during severe droughts, as from 1924 to 1929, they abandoned their territory and took refuge among neighboring tribes or near Government stations.

Meggitt described the Walbiri as living in peace with their neighbors, except for a few reports of warfare, dating back to the previous generation, against two neighboring tribes, the Warramunga and the Waringari. When trespassing into Walbiri country, hunting parties of these adjacent tribes did not restrict their activities to searching for food, which would have been acceptable to the Walbiri, but also included abducting Walbiri women. The Walbiri men would then retaliate with counter raids attempting to retrieve their women. Further incursions into one another's territories led to frequent, and often fatal, fighting.

Around 1900, a conflict arose between the Walbiri and the Waringari over a few water wells near their joint border. The Walbiri attacked, and in a pitched battle which left more than forty dead, defeated the Waringari and took possession of the wells. Since then, the Walbiri have gradually extended their territory by as much as fifteen thousand square miles.[14]

It seems that hunter-gatherers as represented by Australian aboriginal natives are people who mostly live in peace, except when they perceive a need to defend themselves, or impose their essential values on others. They are ready and willing to form raiding parties in order to capture women from neighboring clans, and as a group, they will fight to gain, or at least to keep control over, sources of water. Their strategies in regard to other resources are generally peaceful, mainly because these resources, being geographically spread-out, often mobile and frequently plentiful, are either not worth fighting over or not easily defended.

The raiding for women is an interesting activity, in that it is a form of lethal competition between groups, but also a cryptic form of competition within the group. When going on such a raid, a small group of men start out on a risky business. The group, whose women they are trying to abduct, will fight to their death to protect their women. This can be a fitness cost, while capturing a fertile woman is a fitness benefit. Also, fitness benefits are increased if one of your raiding buddies gets killed and leaves any wives at home, as they are shared among appropriate male members of the group. Any captured women are shared among the surviving raiders. The high mortality among the raiding males, like the twenty-five to thirty percent in the Murngin, does not lower the birthrate of the total population because polygyny, being the accepted mating system, does not leave fertile widows without a new husband for long.

Despite the prevalence of polygyny, other factors in Walbiri society lead to low family sizes. Young men do not marry until they reach a certain status, and many young adolescent women are married to significantly older men, presumably of high status, but with low virility. Unfortunately, neither Warner nor Meggitt give ages of family members, but the sixteen families with children, out of a total of thirty-seven which Meggitt reported, averaged two wives but only 4.7 children. This does indicate a low child to female ratio, for which neither author gives a reasonable explanation. It could be a low birth rate or high infant and child mortality. Both Warner and Meggitt report that infanticide is used by mothers who want to widen the spacing of children, and when twins are born, one is usually killed.

Looking across the vast divide of time and space, from our highly structured society to the Walbiri, we may wonder how a people who claim to be peaceful can fight wars, go raiding for women and commit infanticide. In order to understand such incongruities we must think beyond the box of our own current civilization. People in different countries and regions within a country must contend with different circumstances in order to survive.

Choices for Survival

Regardless of the ways we chose to survive, we are all members of the human family. Watching the two 2007 documentary films, "Flags of Our Fathers" and "Letters from Iwo Jima," powerfully drives home the similarities of the Allies and the Japanese. Each and every one of us is struggling to survive—and to protect the people and resources essential to that goal. We may have arrived at a time when war is no longer a viable way to assure our survival.

If we are unable to eliminate all war, let us hope that the possibility exists for us to adhere to some standard that will reduce the number of wars to those that can be considered totally necessary for our survival on this planet. In President Jimmy Carter's recent book, *Our Endangered Values*, he spells out his vision of the crucial components of a just war, which is a short summary of the original analysis of the topic by Welser[15] and more recently discussed by Atack.[16]

A Just or Unjust War?

A just war should only be waged as a last resort. All non-violent options must be exhausted before war is even considered. Alternatives almost always exist for negotiation and compromise, such as those established by the United Nations and other organizations created to help resolve international crises. Even when no obvious alternative is in place, a time-out period should be declared to explore and develop new options. This certainly applies to developing new sources of energy, conserving water and other critical natural resources.[17]

Weapons used in war must discriminate between combatants and non-combatants. During World War I, our previously held codes regarding this aspect of war went by the board, due to our taking war to the skies and to other technological developments in weaponry that killed masses of non-combatants. Aerial bombardment became a conscious strategy on both sides, leading to the term "terror bombing," for even with some degree of accuracy, bombing always causes great collateral damage.[17]

The degree of violence used in war must be proportional to the injury suffered. For instance, the current war in Iraq cannot be considered "just" by this criterion, because it has caused vastly more destruction and death on both sides than was prevented by the deposing of Iraq's ruthless dictator and his government. There is no possible way that this war was in reality a result of the 9/11 terrorist attack on the United States. It was linked through propaganda to convince a nation that an unjustified declaration of war could be justified. As a result, the United States has become less respected and less safe than ever in its history. International terrorists attacks have more than tripled since the beginning of this war and many more lives have been lost than were lost in the 9/11 terrorist attack.[17]

To declare war, a government must have legitimate authority. In any society, people will hold different opinions on the severity of the problems of war and will have different levels of outrage about war and the willingness to fight in one. The more bellicose members of any society may try to convince other citizens that war is called for, while others may argue for restraint. Through politics and propaganda, governing leaders may try to convince the populace of its point of view. But despite the first three words of the American Constitution, "*We the people*," when a modern state decides to go to war, it is rarely "we" who have weighed all the evidence, discussed possible alternatives, and even participated in the final decision. This is customarily done by a group of political leaders, at best elected by the people, who are often unaware of political leaders' agendas.[17]

The peace to be established must be a clear improvement over what exists. Because wars destabilize regions and governments, a clear and realistic plan must be developed to stabilize what has been destroyed and to establish peace. There is no way to guarantee lasting peace, however, there is a way for both sides to work together to rectify what damage has been done, as the boys in the Robbers Cave experiment were able to do. For this to happen, regardless of the amount of animosity generated during a conflict, those involved must be willing to let go of hate and hold a reality-based vision of better times between themselves and their former enemies.[17]

President Carter writes from the twenty-first century, during which there is an increasing acceptance that we must become more willing to negotiate and compromise to avoid the consequences of possible future wars. At this time, it is also agreed within most civilized countries, that genocide is the most reprehensible of all wars—indeed, a crime against humanity, but it continues as we write. The total annihilation of an entire people was a relatively frequent occurrence in previous epochs, and there is no reason to believe, if we continue to view war as inevitable, that it may become even more pervasive due to the potential use of nuclear and biological warfare.

In 146 B.C., the Carthaginians were wiped out by Rome in the last of the Punic Wars, and closer to modern times, structured native nations in the Americas were dismantled after violent contact with European invaders. As to the future, some Palestinian factions aim at the utter destruction of Israel and anti-Semitism is again raising its ugly head in several other countries. Also, on a larger scale, if the combined threats of global warming and the end of cheap energy are not well managed by mankind over the next decades, we could easily see wars that will once again destroy entire nations—unless we begin to think differently, which requires understanding our past and being able to envision a different future.

Six Million Years of Human Nature

At every stage of our evolutionary past for which we have evidence, we see certain aspects of human behavior and social structure that tell us something about the origins and development of our nature and how it is expressed under different circumstances.

When we examine the branches of our evolutionary tree, we discover that the chimpanzee is not one of our direct-line ancestors, but we share a common ancestor with this ape, who was in all probability much more like the chimpanzee than like us. It is reasonable to assume that our ancestors of six million years ago rarely fought lethal fights within their community, keeping their deadly canines for the battlefield. Both archeological evidence and anthropological studies of early twentieth century huntergatherers and tribal agriculturalists show the same phenomenon.

From the Australian aborigines to the fierce Sudanese Murle, we find socially established rules, practices and beliefs that limit and regulate violence within sovereign communities, while having well developed strategies for waging no-holds-barred war with neighboring groups. In our twenty-first century, we have clear laws regarding violence within our sovereign countries, limited restrictions on violence between countries at peace, and yet, despite agreed upon treaties, like the Geneva Convention, once at war, the rules are quickly downgraded.

An interesting aspect of the externalization of violence is the practice of capturing women to be taken home as extra wives. Chimpanzees do it, hunter-gatherers do it, pre-historic tribal communities did it—and even the Bible is full of it. In the Old Testament book of Numbers, Chapter 31, Verse 18, we read that Moses, after ordering his troops to kill all men as well as all women who were not virgins, concluded with: "...all the women children who have not known a man by lying with him, keep alive for yourselves." The book of Deuteronomy gives one of the first references to this practice, with a "civilized" twist to the treatment of the captured females. When the sons of Israel gained a new territory through battle, they were allowed to take the choice females as wives, as long as they gave the women a month to grieve their dead husbands.[18]

As with the Walbiri and their neighboring tribes, the reason that the practice of capturing women results in externalization of violence is that polygyny in a social species with a fifty-fifty sex ratio tends to lead to intra-community tensions, which can easily lead to violence because some men have several wives and others have none. Greed and jealously remain aspects of our nature and are prone to promote violence throughout our lives. Allowing the importation of forcibly captured foreign women in combination with the battlefield deaths of men can help mitigate this source of intra-community violence. It is not too far-fetched to see the voluntary import of war brides from a war-ravaged county as a civilized form of the same phenomenon. It is becoming a totally accepted practice for men who have not been able to find a suitable wife in their own country to seek partners from third world countries through legitimate organizations. The woman often gains an opportunity for a more comfortable life, more education and a higher standard of living for herself and any children she may have.

In 1985, I (Dolf) became acquainted with an Australian man while we shared a hospital room in Cairns. His much younger wife and two teenage children frequently visited Patrick. While they were not present, he told me their story. He had worked for decades as a sailor in the merchant marine and had saved his money until he retired at age sixty-five. He then placed an advertisement describing himself and asking for a good Catholic middleclass woman of child-bearing age in a Bogotá newspaper.

He received several replies from interested women and visited their families before making a choice. After marrying the woman of his choice, they settled in Cairns, where they run a successful motel together. Their two children are doing well in school; one excels in the sciences, the other in oriental languages with his sites set on a career in international trade. On one of her subsequent visits, she told me that at thirty years of age, she did not consider herself pretty and thought her chances for a good marriage in Columbia were low, and that Patrick, despite being more than twice her age, had proven to be a wonderful husband.

During these past six million years, we have fought wars over resources, and in a strictly bio-scientific sense, males of our species relate subconsciously to women as a resource. The other essential resources over which we tend to fight are land, water, oil and any material we consider necessary to maintain our economic standard. Without women and material resources, a man would have no chance to reproduce, hence, no fitness. If men fought one another over resources within the group, group cohesion would collapse, leaving the individual members easy victims of inter-group aggression. Instead, we have evolved a complex blend of social cooperation and limited competition within the group, which maximizes reproductive success for those men who most effectively defend the group and help maintain the group's dominance over its neighbors, especially if they capture women when raiding other groups.

Resources have always been considered fair game, to be acquired from beyond the community, via violent means, if necessary. Those societies that were successful in this respect grew in population size, endowing its members with higher reproductive output than that of the average member. The essence of the above argument is that to have an internally peaceful society without violent competition among its members, our ancestors considered it desirable to provide the citizens with the necessary resources to allow a family structure, which would result in population growth, but which inevitably would lead to war over resources as population expansion leads to diminishing resources.

Understanding this, we must create a new model if we are to maintain life on our planet for future generations.

The fact that we have survived at all attests to our ability to change our feelings, thoughts and behaviors in response to our needs and experiences. When we step back for another brief look at the more primitive relations among chimpanzee communities, we have seen that, as the populations of two neighboring communities grow and resources become scarce, a war of attrition develops, which will inevitably lead to the extermination of one of the groups. But, we are not chimpanzees; we have the ability to grasp future consequences of present behavior. We can assess a difficult situation and have the capacity to change what must be changed to create new opportunities for survival. We believe that we can change our own behavior in such a manner as to adopt a realistic strategy for the grand solution: a world without war.

References:

Q. Clemenceau, George. *Power Quotes*. In Baker, Daniel B. (Editor) *Power Quotes*. USA: Visible Ink Press for Barnes and Noble Books, 1992.

1. Petrequin, Harry. National War College papers.

2. Owen, Wilfred. Preface to Collected Works. New York: New Directions Book, 1963.

3. Bradley, James. *Flags of Our Fathers*. New York: Random House. 2000.

4. In Baker, Daniel B. (Editor) USA: Visible Ink Press for Barnes and Noble Books, 1992.

5. Hedges, Chris. *War is a Force that Gives us Meaning*. New York: Public Affairs Press of Perseus Books Group, 2002.

6. Sherif, M. et al. *The Robbers Cave Experiment, Intergroup Conflict and Cooperation*. Wesleyan U. P.,1988.

7. Golding, W. G. *Lord of the Flies*. United Kingdom: Faber and Faber, 1954.

8. Fry, Douglas P. *The Human Potential for Peace: An Anthropological Challenge to Assumptions about War and Violence*. Oxford: Oxford University Press, 2006.

9. Keeley, L. H. *War before Civilization, The Myth of the Peaceful Savage*. Oxford University Press, 1996.

10. Gregor, T. *Uneasy Peace: Intertribal Relations in Brazil's Upper Xingu*. In In Haas, J (Ed.) *The Anthropology of War*. Cambridge University Press, 1990.

11. Le Blanc, S. A. *Constant Battles, Why We Fight*. St. Martin's Griffin. 2003.

12. Jansen, M. (Ed.) *Warrior Rule in Japan*. Cambridge, England: Cambridge University Press, 1995.

13. Warner, W. L. *A Black Civilization. A Study of an Australian Tribe*. Harper and Row, 1958.

14. Meggitt, M.J. *Desert People, A Study of the Walbiri Aborigines of Central Australia*. Un. of Chicago Press, 1965.

15. Welzer. M. *Just and Unjust Wars: A Moral Argument with Historical Illustrations*. 2nd Edition. New York: Harper Collins, 1992.

16. Atack, I. *The Ethics of Peace and War*. Edinburgh: Edinburgh Press, 2005.

17. Carter, James E., *Our Endangered Values*. New York: Simon and Schuster, 2005.

18. *Holy Bible*

CHAPTER

9

Shaping a Sustainable and Peaceful Future

*The future of everything we have accomplished since our intelligence
evolved will depend on the wisdom of our actions over the next few years.
Like all creatures, humans have made their way in the world so far
by trial and error; unlike other creatures, we have a presence so
colossal that error is a luxury we can no longer afford. The world has
grown too small to forgive us any big mistakes.*
~ Ronald Wright

The twenty-first century has begun with a grim warning from the
scientific community: Earth's ability to meet the needs of our growing
population is crumbling. Quoted above, the historical philosopher,
Ronald Wright, in his popular book, *A Short History of Progress*, adds

to the warning of the scientific community by emphasizing that citizens of a world with a fragile and unraveling biosphere, resource shortages and occupied by competing nations with nuclear weapons must wake up, stop wasting words and start taking immediate action to survive.[1]

Our life here on Earth cannot be sustained unless we control population growth by safe and ethical means, manage and preserve our natural resources and treat others with respect and compassion. In other words, we must soften our human footprint on Earth. The longer we postpone initiating this vital work, the closer to impossible it will become to shape a sustainable and peaceful future. Our work requires a willingness to change—to use our evolved intelligence to implement action that will halt the rapidly growing threat to the survival of our civilization.

Our Present Footprint

It may seem incongruous to think about the collapse of our global civilization while we experience rapid economic growth, impressive advances in health and nutrition and expanding commitments to peace. Yet, the voice of history screams loudly that local and regional civilizations have nearly always collapsed shortly after reaching their peak. This is no mere coincidence. As civilizations develop, they tend to overuse available resources, destroy environments faster than they can be recovered and allow population growth beyond their region's sustainable capacity. When such problems become apparent, they are often dealt with through "quick fixes," which can offer relief on the short run, but create greater problems down the road. Eventually, these problems mount up to form such a precarious balance that a single bad harvest, a military defeat or a natural disaster can collapse the entire civilization. This phenomenon can be seen proverbially as the footprint of the human population on the Earth. Our current global footprint is bringing our future on Earth to the brink.

In his most recent book, *Plan B 3.0: Mobilizing to Save Civilization*, Lester R. Brown, President of the Earth Policy Institute, vividly describes the resource trends, which without major reversals, will destroy our civilization. He explains that the annual increase in global population has already exceeded Earth's ability to supply adequate resources for us to survive longer than a few more decades. This harsh reality is the result of the degradation and loss of arable land, deforestation, depletion of water resources, pollution of land, rivers and oceans, and of global warming.

The roots of this dilemma lie in the mega-growth of the global economy and human population during the past century. The world economy has expanded twenty-fold since 1900, and the world population has quadrupled. A United Nations sponsored four-year study of the world's twenty-four primary ecosystems by 1,360 scientists, published at the beginning of the twenty-first century, reported that fifteen of these ecosystems are being pushed beyond their limits of renewal.[2]

Creating a Sustainable Future

The future is not some place to which we will go, but a place that each of us is creating. There is a rapidly growing gap between what human nature has programmed us to do and what our rational minds know is necessary for survival. What we do to close that gap will determine our future. Previous chapters have shown that our willingness to wage war increases when it becomes obvious that we will soon suffer a shortage of essential resources. Over a long evolutionary history, human beings have nearly always preferred to fight than to starve. This deeply ingrained trait will propel us toward the fate of the dinosaurs, unless we create a sustainable balance between population and available resources. We must give up our denial of this harrowing reality.

Another problem is that contrary to well established evidence, many of us still think of nature as an unlimited source of whatever we can extract out of it. We need to understand that for nature to nurture us, we must nurture nature in return. As human nature has evolved, we have not developed appropriate responses to the hazards we have created for Mother Nature. Biologist E. O. Wilson has declared that "the most ravaging monster upon the land is population growth and that until we conquer its presence, sustainability of our planet is but a fragile theoretical construct."[3]

Wolves under the Midnight Sun

For possible enlightenment toward taming this monster, we turn again to the animal world. In previous chapters we have introduced some of our favorite animals to point out similarities and differences between them and us, and to illustrate some concepts that apply equally to the entire animal kingdom—including humankind. We have described competition and cooperation in muskoxen, play and behavioral rules in sea otters, infanticide in owls and polar bears, and socio-culturally-driven natural selection in monkeys and chimpanzees. We will now introduce the white Arctic wolf (the subspecies *Canis lulpus arctos*) which has evolved a complex way of dealing with competition for rare and unpredictable resources. We believe that humankind can learn something significant from this species, in the area of resource management—and in the realms of love and war.

While doing research on Banks Island in the Canadian Arctic, my students, colleagues and I (Dolf) frequently encountered these magnificent predators. Most of our encounters involved seeing one or a few wolves at a fair distance—usually in the act of running away. Their skittishness is largely due to the fact that the local Banks Islanders rarely come as far north as our research area, but when they do, they arrive on ATVs or snowmobiles and shoot any wolves they come across. Therefore, the local wolves tended to stay well away from our camp. However, toward the end of our stay, the wolves seemed to recognize us as somehow different and to be less afraid of our presence. Then, one night late in the summer, a student whispering insistently at my tent's entrance awakened me: "Dolf, the wolves are here! Come out and see!"

I dragged myself out of my sleeping bag, staggered out of the tent into the still-bright midnight sun and found myself standing barefoot in some freshly fallen snow. Barely thirty yards away, stood three fully-grown wolves. They looked intently at us; we looked back at them. For a few moments, no one moved, until two of the wolves turned around and slowly slunk away. But the largest stood his ground; he seemed to stand as high on his paws as he could. Side-lit by the low northern sun, his white fur wafting slightly in the breeze, he stared at us with what seemed like a mix of pride and resentment; who were we to tread on his domain?

I was so absorbed by this silent confrontation that I forgot the frigid cold; time seemed to stop while this wolf and we stood perfectly still, staring at one another. Then suddenly, the wolf lowered his head, shook it slightly, made a low growling noise, and then threw his head back and filled the valley with a sonorous, spine-chilling howl that echoed off the surrounding hills. As the last echoes died into silence, he turned and trotted off with a high-in-the-saddle demeanor. Later, back in my sleeping bag, I lay awake shivering, considering how this wild being might fit into my disciplined, scientific vision of the world shared by man and beast.

The Life Strategy of the Wolf

The wolf is a large predator at the narrow top of the food chain. Because of this position in the ecosystem, wolves are rare, and when their prey species decline, wolves go hungry. It is a general ecological rule, with few exceptions, that predators are ultimately controlled by resource shortage, while herbivores are controlled by their predators. In many species, when individuals face resource shortages, they will fight among themselves for access to limited resources. Most animals prefer to fight than to suffer hunger with its associated illnesses and starvation. The wolf is no exception. Like most social animals, wolves limit violence within the pack, while fighting between packs (or between individuals of different packs) can be vicious and deadly. Natural selection has taken an interesting turn in the evolution of this animal, reducing inter-pack violence as well.

Wolf packs have a complex social structure and associated reproductive strategy. Like the chimpanzee community, the wolf pack has two dominance hierarchies, one for each sex. Males are always dominant over females, but here the similarity ends. In a small wolf pack, like the one studied for ten years by biologist Dr. David Mech on Canada's Ellesmere Island, only the dominant female produces a litter of pups in any one year, even when there are two or three mature females in the pack. But, all the members of the pack, females, males and juveniles, help feed and take care of the pups. At first consideration, this seems a rare example of altruism on the part of pack members who are not the parents, as well as a puzzling case of group-based birth control.

Let's look first at the apparent altruism. Long-term breeding observations by Mech and other wolf biologists have shown that inbreeding in wolf packs is not rare, which means that by taking care of any given year's pups, whether one is the parent or not, one's inclusive fitness increases relative to the average fitness in the total wolf population. While this is a powerful argument, it does not explain all aspects of the situation. Some immature wolves leave the parental pack to join or help form other packs. When not related to the other pack members and not a parent of a new litter, these animals still assist with the rearing of the pups. This is what we will call "negotiated altruism." While wolves obviously do not negotiate responsibilities or sign notarized contracts, nonetheless, a new pack member gains a great deal of security by belonging to the pack for many years and can reasonably expect to gain an opportunity to become a breeder in the future.

Another initially puzzling aspect of wolf behavior is the acceptance by lower-ranking females, and probably males, to refrain from breeding when the pack can be expected to raise just one litter per year. While every related pack member gains some fitness when a litter is raised successfully, the actual parents gain considerably more. A more careful analysis suggests an explanation. The dominance hierarchy is very rigorous in both sexes, and during the brief breeding season, the higher-ranking individuals constantly harass the lower-ranking ones when they attempt to mate. This can cause high levels of stress, which interferes with normal reproductive physiology and usually prevents a successful pregnancy, even when mating has taken place.[4,5]

While stress may be a proximal cause of a lowered reproductive rate, it is not the ultimate cause. Natural selection would result in a gradual loss of stress in harassed wolves, but only if wolves under less stress did get pregnant and raised their pups successfully. In the real world they probably would not, because the sparse and harsh high arctic environment would not deliver enough resources to raise more than one litter. Hence, natural selection favors the stress effect on lower-ranking wolves when harassed by higher-ranking ones. On the average, the lower-ranking stressed individuals do better by earning an increment in their inclusive fitness than by producing doomed litters, while also endangering the survival of the litter produced by the top-ranking female.

The ultimate cause of wolves existing in packs with strongly enforced dominance hierarchies resulting in a reduced pack reproductive rate is not the behavior of the high-ranking individual members. Nor is it the resulting stress experienced by the lower ranking members. Instead, it is the limit set by the resources of the environment. This conclusion may be challenged by some biologists who see a world filled with prey items; however, we believe that the wolves are capable of assessing not just resources available at any given time, but also what can be expected to be present over the period needed to successfully raise a litter. They have evolved the skill to assess their environment, and when conditions indicate a probable shortage of prey, they have adopted a life strategy that successfully limits reproduction.

Worthy Wolf Lessons

We realize that the wolf's life strategy, including its built-in birth control mechanism, is not thought out rationally by the alpha individuals and imposed upon the pack. The strategy simply evolved as an effective way to maximize the fitness of most of the pack-members for long periods of time. Due to the lower birth rate, wolves also have a lower mortality rate. Packs produce far fewer pups than they could, but willingly spend effort and resources to bring a higher proportion of pups to adulthood. This results in a more stable population, both within the pack and for the wolf population as a whole. This stability, in turn will tend to externalize competitive aggression to beyond the pack, while also reducing inter-pack violence.

Under some ecological conditions, it is possible that our hunter-gatherer ancestors practiced a similar strategy. Anthropologist Dr. Keith Otterbein poses the intriguing possibility that early in our human evolution, as we became hunter-gatherers on the African savannah, we were highly nomadic and dependent on unpredictable resources. Otterbein believes that until the early beginnings of agriculture, our ancestors were peaceful. However, in view of the prevalence of war among more recent hunter-gatherers when relatively good times declined to harsher conditions with a reduction in available resources, we doubt that earlier hunter-gathers totally avoided war. Nevertheless, a strategy similar to that of the wolves may have prevailed over long periods of our early evolutionary history.[6]

We could learn from the wolf, and perhaps from some of our early ancestors, how to design a version of their life strategy. But, for us to achieve this strategy quickly will be more difficult than it was for the wolf, who over many generations simply responded to natural selection. On the other hand, we have the intellectual and reasoning power to design new life strategies in anticipation of future conditions, as well as the adaptability to rationally overrule many of our evolved urges.

Our Final Hour?

Our world population is now over six billion, and demographers forecast that it will level off to between eight and ten billion later this century. Clearly, ten billion people will not be able to survive on this planet if current economic growth continues with its inevitable increased resource use. We could only sustain that number if all of us were willing to live in dire poverty, but wars over limited resources would almost surely break out. Even seven billion people will not be a sustainable population at an acceptable level of sustenance, education and health.

This harsh reality was recently driven home by Britain's Astronomer Royal, Sir Martin Rees, in his book, *Our Final Hour: A Scientist's Warning: How Terror, Error and Environmental Disaster Threaten Humankind's Future in this Century on Earth*. He warns that the most benign outcome, if we can survive this century without catastrophic reversals, would be a world population significantly lower than at present.[7]

Since it is unlikely that we will be able to increase our resources to a level that will support the anticipated human population by the end of this century, we must seriously think about how to reduce population growth. Whether we can achieve the benign outcome Rees hopes for is questionable, but the un-enforced slow downward spiral of population in some European and other industrialized nations, as well as the partly enforced Chinese one-child policy, are worth studying as possible models.

Taming the Glutton

In the English language, the word *growth* generally conjures up good feelings, despite simultaneous awareness that growth in the wrong place and at the wrong time can be catastrophic. Economists have no problem recognizing that when a fisheries industry outgrows the fish population's ability to replace what has been caught, disaster ensues. The same holds for forests, fresh water reservoirs, and all other renewable resources. We have to recognize that the world's economy is rapidly outgrowing the earth's ability to replenish what we are taking out, as well as its ability to absorb and/or detoxify our waste.

It is a "wonky world" that moans when the annual growth in GDP is less than three percent, while simultaneously worrying about where our future low entropy energy supplies may be found and how we are to stem global warming. Global warming is no longer a theoretical possibility; it is here, gravely threatening our ability to feed our massive and still growing world population, while similarly threatening the survival of many other species.[8]

Simultaneously, discovery of new oil deposits peaked a few decades ago, and production appears to be reaching its peak. There are only a few places on Earth where the extraction of fossil fuels can be increased, and only at rapidly growing cost, both directly in production cost and indirectly in increased release of greenhouse gases. Fossil fuels, metal ores, uranium and many other resources are not renewable. Some are plentiful, but most will become increasingly expensive to extract because the richest deposits have already been exhausted. Finding ways to recycle material resources and develop sustainable alternatives to fossil fuels are time-consuming and expensive, but urgent.[9]

Two strategic approaches could be adopted to tackle this problem. The first is highly risky. It argues that rapid economic growth over the next twenty or thirty years would provide the wealth and technology to make the necessary transition to a sustainable world before it is too late. This approach feeds the glutton of growing economies. The other strategy starves the glutton by playing it safe. It argues for reducing economic growth now, which will give us more time to develop the technologies needed for the transition to sustainability. Unfortunately, the first alternative is preferred by current politicians, as well as by the industrial and business sectors of the developing world. But, it is risky due to the high possibility that we will soon find ourselves with a rapidly expanding world economy, unexpectedly high global warming with associated environmental deterioration, and without adequate wealth and technological advances to provide solutions.

A prime example of this is an ill-conceived effort by the United States to reduce its dependency on imported oil by implementing plans to convert thirty percent of its grain crop into bio-fuel, primarily for private motor vehicles. This massive diversion of food grains to ethanol distilleries has so increased world food prices that the number of hungry and malnourished people, now estimated at 830 million, is expected to increase to 1.25 billion by 2025. This can only increase the probability of civil strife within poorer nations and conflicts between nations. Ironically, implementing such a policy which makes already harsh living conditions among poor nations even worse, can only foster increased resentment—as well as reinforce whatever inclination those affected might have to undertake retribution, including terrorist acts. This is an egregious oversight while politicians claim to be conducting a "War on Terror."

Other nations have managed to produce bio-fuels from non-edible grasses, pressed sugar cane, corn stalks and other plants. To divert edible grains for this purpose in the face of massive hunger and malnutrition among the world's population has been classified as a "crime against humanity" by the UN World Food Program.[10]

It is high time that our economists and politicians start to consider long-term side effects of short-term solutions when trying to build a stable economy. Once we have a sustainable economy, we can again aim for economic growth, but only when it does not disturb the sustainability equilibrium. For now, we must tackle the problem of how to structure an economy that can guide us through a lengthy soft landing towards a lower level of consumption in tandem with population reduction.

Swords or Ploughshares

In the early 1970s, shortly after the publication of the influential report, *The Limits to Growth: Club of Rome Project on the Predicament of Mankind*[11], I (Dolf) attended a meeting of economists, scientists and engineers to discuss the future of humankind. The economists were highly critical of the report, considered the authors ill-informed doomsayers, while they, the economists, were confidently optimistic about the future.

By contrast, the scientists and engineers were impressed by the report, despite some significant criticisms, because they had serious doubts about our ability to keep expanding our footprint on Earth without some major collapse in the near future.

What I found most disturbing was that the economists' confidence was solidly based on their conviction that we, the scientists and engineers, would come up with the necessary technological solutions to the problems. The green revolution, genetic engineering, better pesticides, off-shore oil drilling, better methods for food distribution and other advances have, in retrospect, proven the economists right—at least for the past quarter century. But, for how much longer?

Both the total amount of food per person and the quality of our food has increased over the past century and will probably continue to do so for the next few decades. Yet, there is a limit to what even the most sophisticated technological advances can accomplish. In the short term, there will be further increases in the use of fertilizers, introductions of new species of food-plants, better pest control and genetic manipulation of food-plants, both through breeding and genetic engineering. Other high-tech methods, such as GPS-driven manipulation of patch-by-patch control over irrigation, improved fertilizer use and more effective pesticide application, will also make a difference. Even better, ecology-based biological management of agricultural systems by subtly manipulating interactions between crop plants, pests and beneficials (organisms that control the pests) would allow major advances in food production without unwanted side effects.

Biological systems can be pushed only so far. The closer we come to that biological limit, the greater the cost for each increment of increased food production. Over the same time period, energy costs will rise dramatically, water shortages will increase and the equipment needed for agriculture will become more expensive. The most threatening problems are those that are slowly corroding agricultural production, and they will become more severe. These problems include: the gradual deterioration of the land on which we grow our food, desertification due to overuse and climate change, loss of organic matter, erosion, accumulation of salts, and urban sprawl. Such problems will accelerate at a faster rate once we run out of technological trickery to con nature into increasing agricultural output. Already, world grain production has fallen short of consumption in seven of the last eight years. World grain stocks are now at their lowest in thirty-four years. These facts can no longer be denied—nor ignored.

Furthermore, agriculture is not an isolated industry. Irrigation often uses water that is also needed in cities; the use of pesticides in agriculture has negative effects on natural stability and biodiversity; and expanding agricultural land use inevitably eats into natural areas. Fertilizer use causes runoff into rivers resulting in high levels of eutrofication, which too rapidly increases the growth of algae, thus depleting oxygen for the fish population, and thereby destroying the aquatic ecosystems. These and many other problems make it imperative that we focus more on those aspects of science and technology that can push our potential food production towards sustainability, not only within the focused framework of agriculture, but for all other aspects of existence in the biosphere.[12]

To illustrate the problem of land deterioration, let me dig into my (Dolf's) diary and pull out a paragraph from my January 30, 2006 entry from Malaysian Borneo:

> *This ride through the interior of Borneo was a very sobering eye-opener. For some 100 km, we drove through virtually nothing but oil palm plantations on both sides of the road for as far as we could see. There were old established ones, young ones, recently planted ones, and ones that were recently cut and were being prepared for re-planting. The realization that all of this land was primary rainforest half a century ago was bone chilling. The fact that another area of rainforest, twice as vast, is slated to be cleared for oil palm over the next few years is even more disturbing. The scientific forecast is that despite massive use of fertilizer and pesticides, the soils of the entire area will become so degraded that in another 50-100 years nothing of commercial value will be able to grow there. If such ex-plantations are simply abandoned, it will take centuries of scrubland to rebuild the soils and eventually re-establish a rainforest. Oil palm cultivation has to be one of the most unsustainable forms of agriculture.*

What we saw illustrates three serious problems with current agricultural strategies, especially in the tropical developing world. While the transfer from rainforest to plantations provides the local region (in this case Malaysia) with a valuable source of oil for human consumption and industrial use, it destroys much of the rainforest, causes soil degradation and erosion. Locally, the practice is seen as a basis for economic and human population growth, although the long-term effects are destructive. While we have chosen agriculture as the focal example of our need for improved technology, wiser uses of technology are equally important for other aspects of establishing and maintaining a sustainable economy.

One of our most serious problems is the amount of scientific and technological wizardry we spend on the military. In 1795, U.S. President James Madison wrote:

Of all the enemies to public liberty, war is the most dreaded because it compromises and develops the germ of every other. War is the parent of armies; from these proceed debts and higher taxes, instruments for bringing the many under the domination of a few. No nation can preserve its freedom in the midst of warfare.[13]

Global military spending and arms trade during the twentieth century total over one trillion dollars per year. Meanwhile, the United Nations and all of its agencies spend only a tiny fraction of that amount—twenty billion (two percent of one trillion)—to help provide food, education and health care for those who live in poverty on this planet. Recently, funding for many of these important programs has been cut, due to the financial drain of ongoing wars.[14] We consider these policies another "crime against humanity." Maintaining national security is important, but the way most countries spend for military purposes is not getting the job done. Terrorism and ill-directed fear are at an all-time high the world over. If more of these resources could be allocated to solving our mounting problems of poverty, ignorance and environmental destruction, we could come closer to developing a sustainable future. The best we can hope for is that the leading militarized countries will agree to a major reduction in military spending and start spending more on education, science and appropriate technology. This holds especially for nuclear research, if applied to the search for safe, clean energy instead of for weapons; this would be a step toward making war obsolete.

Brutalizing Mother Nature

There is an interesting progression from so-called primitive religions, which worshipped nature, to the more advanced monotheistic ones, that has gradually separated the spiritual from the natural. If paradise were still envisioned as an old-growth forest with a stable ecosystem and high biodiversity including snakes, apple trees and a sparse human population instead of pearly gates leading to something like a deified shopping mall, Earth might be in better condition today.

In industrialized countries, it is difficult to convince governments that the protection of wild areas is of major importance to our future. Protecting natural areas, such as the rainforests in developing nations, is proving to be a very hard sell because oil is the prized commodity. In Borneo, the Danum Valley Conservation Area has been set aside for the protection of the rainforest and as a home for major wild animals, such as the Bornean rhino, the Bornean elephant and the orangutan. The area of the reserve is not sufficient to maintain healthy populations of these large animals, but the much larger area surrounding the reserve is, for now, adequately suited for the wildlife.

There is a conflict of interests between those who want to replace that surrounding land with oil palm plantations and those who argue that in the long run, management of the existing forest with reduced-impact logging methods not only protects the forest and its wildlife, but will add more real value to the life of our planet. It may be necessary for Western countries to offer financial incentives to tip the balance in favor of preserving the forest in much of the tropics, which will help to sustain the entire world.

In Indonesia destruction of many thousands of square miles of rainforest, especially of those forests that grow on millennia-old peat swamps is an even greater problem. When such forests are cleared and drained, between 1500 and 2250 tons of CO_2 per acre are released over the next twenty-five years due to the burning of cut vegetation and the subsequent decay of the peat. The palm oil that will be harvested on one of those acres for the production of bio-diesel for export to Europe will only reduce Europe's CO2 emissions by approximately sixty tons over the same period. Europe (and the entire world) would be way ahead if they paid Indonesia to leave the peat-land rainforest as it is, and burnt fossil fuel instead.[15]

In his book, *Forests in a Full World*, George Woodwell, Director of the Woods Hole Research Center in Massachusetts, decries the rate at which tropical forests are being cut. His most optimistic prediction places the destruction of the last unprotected tropical forest at approximately one hundred years from now—his most pessimistic prediction at forty years.[16] We are one of millions of species that share a long history of life under harsh natural conditions. All the living matter on Earth has been, and is still being, shaped by the conditions of the biosphere; in turn, that living matter has shaped and continues to shape that biosphere. The oxygen in our atmosphere is produced by plants and algae. Most of the greenhouse gases, such as carbon dioxide and methane, are also mainly of biological origin. Our soils are produced by a multitude of living organisms and an infinite number of interactions among them, many of which result in death or create new life, thereby maintaining the biosphere's remarkably stable equilibrium. Similarly, our current climate, which provides for an abundance of liquid water, and hence, the existence of a living biosphere, is also maintained by that biosphere at a stable equilibrium.

So far, global warming has merely shifted the equilibrium point to a somewhat higher temperature with predictable and costly, but not yet, catastrophic consequences. There is, however, a growing risk that our unnaturally high release of industrial greenhouse gases will push that equilibrium beyond its current stable zone into a new stable zone with considerably higher temperatures, which will have major climatic catastrophic results for the entire biosphere. The greenhouse gases that spew from the United States, Europe, and China are endangering lives in poorer, mostly tropical countries, which emit very little carbon, but they will suffer from the way our carbon emissions will affect their climate.[17] Such rapid shifts from one negative feedback-based stable zone to another are nearly impossible to predict with any accuracy;

however, ignoring them is analogous to allowing young children to play with matches.

We can minimize the risk of Earth's climate and biosphere suffering a major collapse by keeping a large proportion of the total land mass covered by forests and the oceans adequately stocked with algae. As long as the natural cycle of photosynthesis and respiration remains strong, our climate will be more likely to remain within its natural stable range that will allow us to adopt an appropriate, sustainable life strategy. If we inflict more serious injury upon our already battered Mother Nature, she could die on us, taking her children with her.

We, who are lucky enough to be alive on this planet today, have a critical obligation not to merely avoid further damaging Mother Nature, but to nurture her and to maintain as high a biodiversity as possible. Beyond moral issues, there are two good economic reasons for doing so. First, ecosystem stability depends on complex-interactive trophic webs, which in turn depend on a large number of co-evolved species. One of the problems in agriculture is that our crops, especially ones grown as monocultures, such as corn and rice, are not stable, and can suffer serious losses due to pests and weeds.

Preventing such losses with pesticides and weed-killers is costly and environmentally destructive. Good biological management in agriculture depends on learning to live with a low level of pests, weeds and other species, which form a semi-stable ecosystem long enough to bring an adequate crop to harvest. The other reason to maintain high biodiversity is that both species diversity and genetic diversity are essential for us to adapt food and medication production to changing conditions as the climate and our life strategies change. Without the raw material, neither selection nor genetic engineering can progress.

An example of the kind of problem we may face is that the current varieties of banana are non-seed producing, and therefore are a-sexually reproduced by taking cuttings. This inevitably leads to genetic impoverishment. Large monocultures of banana of near homogeneous genetic make-up are highly susceptible to diseases and pests, as the disease organisms and pest species evolve much faster than the a-sexual bananas. This cannot be rectified by breeding, which until recently would have been a serious predicament. Genetic engineering can use genes from wild species of banana that still flourish in South-east Asian rain forests to introduce such traits as disease resistance in commercially productive banana varieties.

Venus versus Mars

The vast majority of humans throughout the ages have claimed to desire peace. We have professed our love for Venus, the Goddess of Love, but continue to feed Mars, the God of War; not only when an imminent threat is staring us in the face, but also between periods of conflict to keep abreast with the armaments of potential enemies. The God of War may be doing his dirty work on a relatively small scale, but the small wars currently in progress are like the rumblings of a dormant volcano. When major shortages of essential resources come, the armories will be unlocked. Mars will be knocking at our doors, offering us his services.

We have no totally reliable evidence that there was ever life on Mars, but there is speculation that there could have been some form of living matter present in the distant past. If that were the case, we know that the very thin atmosphere presently surrounding Mars could not support a thriving biosphere. It is not unimaginable that in the distant future, whatever and wherever intelligent life forms may exist will be trying to analyze Earth from a similar perspective.

The dilemma would be to determine whether the biosphere of our planet had been destroyed by the carelessness and lack of foresight of the inhabitants, or had there been a nuclear holocaust that caused its demise. This, of course, is all conjecture, but what we do know is that as long as nuclear weapons exist, the threat of nuclear war hangs over our heads. In 1997, the Canberra Commission published a report on its deliberations regarding the future of nuclear weapons, which included the following assertion: "The proposition that nuclear weapons can be retained in perpetuity and never used, accidentally or by decision, defies credibility."[18]

We do not believe that any of the major powers in the world would deliberately initiate a nuclear war with another major power, because government leaders would realize the retaliation from such an act would end our civilization. Nevertheless, as long as nations are building and storing nuclear weapons, we face the risk that some crisis could escalate into the use of these weapons. There is also a risk that through technological or computer malfunction, there could be a catastrophic nuclear mishap. The major purpose of building a nuclear arsenal is that of threatening the enemy, however as conflicts are allowed to escalate, we seem to lose control of how they might end. If we allow the continuation of conflicts without using every non-violent means of settling them, Earth could easily end up as dead a planet as Mars.[19]

Statistically speaking, any event that has even a small probability of occurring, unless the probability diminishes over time, will eventually occur. Therefore, one of our generation's greatest challenges is to diminish war, or nuclear war will destroy the human species. One of the most powerful lines in the 1985 PBS National Television Series, *War*, written and narrated by Gwynne Dyer, states that we are living in the Indian summer of human history, with nothing to look forward to but the "nuclear winter," that will close our account.[20]

To minimize the threat of a nuclear holocaust, we must commit to a vision of a world in which war will be considered some bizarre behavior from history, ghastly, scarcely imaginable—and totally unnecessary. For this vision to take hold, we must struggle toward a deep and honest acceptance of our true nature, of the abiding presence of our two internal wolves, one, competitive and the other, cooperative, and make wise decisions about which one to feed.

In earlier chapters we discussed the sources and causes of war. Although family feuds, religious squabbles, the greed of despots, and other such proximate causes, are frequently put forth by people, the ultimate cause of war is nearly always limited resources or the fear that resources will become limited. This fear combines with the belief that "the others" are the source of threat, and we deem them "enemies." Once, this division is experienced, we become angry at "the enemy." Our anger leads to violence, and violence leads to war. This is why animals fight—including the human animal. We have evolved this way because, in the past, we were more likely to survive if we obtained greater supplies of resources, even if we had to take the deadly risk of fighting our neighbors for them.

Minor wars and their aftermath, despite all the death and destruction they cause, could still create the selective environment to further this trait. The existence of nuclear weapons in arsenals the world over, with the itchy fingers of generals, presidents and religious maniacs hovering above the buttons, have made this evolved trait a seriously maladaptive one. Reaching for our weapons as soon as we feel threatened must be overruled by reason, for only then will humankind no longer feel threatened, or in need of threatening others. We must realize that we can learn from others, and that working with others toward solutions can enhance life for all of us. Until we have established such a rationally based global society, we will not be free from the primal ignorance and fears that drive us to war.

Our world is blessed with people who work tirelessly for peace. During 2007, an international panel of eminent intellectuals deliberated on how best to prevent war in the future. Their report, "Civil Paths to Peace," tackles a broad spectrum of social and political ills that must be addressed, including poverty, inequality, problems in education, the biased media and lack of political participation, if we are to establish a more peaceful world. The emphasis throughout the report is on improving human relations, especially developing more respect and understanding.

Well thought out arguments are presented for the need to provide all members of human society with real opportunities to participate in political and economic life. The report asserts that if we give everyone a voice that will be heard and respected, much violence could be prevented.[21] This report is representative of many peace plans that speak convincingly to our hearts. It is appealing to think that if we embraced all the people of the world as our brothers and sisters, war would simply vanish.

This belief is not flawed in any essential aspect, but it is incomplete. It ignores our primal nature, our long-evolved, deeply ingrained trait to choose fighting over starving. The only way through this dilemma is to claim the potential of our higher consciousness to change our course—to move beyond evolution, or someday, a more rational life form will be examining our extinct biosphere.

Overcoming Social and Cultural Obstacles to Peace

Included in serious obstacles to achieving a sustainable and more peaceful society are those social and cultural aspects of our identity based on irrational ideologies. Beliefs propagated by certain doctrinaire and political movements, religions and commercial interest groups, and some well-intentioned, but misguided, charitable organizations often have sown seeds of ignorance and fear between groups.

In a free, secular and open society, the freedom of individuals to hold personal religious beliefs and the right to belong to a group of similar believers is sacrosanct and should be respected. And yet, many societies have interpreted the right of religious freedom to include the right to practice various non-spiritual aspects of their religious culture. When that practice involves the wearing of a certain headgear or the eating of a specific food—normal rights of any citizen—there should be no problem. On the other hand, when it involves a behavior or an item that could interfere with, or even hurt, other citizens, it should be questioned with an open mind on both sides. The failure to make the above distinction has caused problems between established members of secular societies and immigrants coming from more traditional, non-secular backgrounds.

The same confusion in traditional societies has led to the opposite problem, namely that the established societies expect that they should have the right to practice the non-spiritual, cultural aspects of their religion and run afoul of conflicting practices of other equally inflexible religious groups. In more homogeneous traditional societies, the non-spiritual aspects of the dominant religion are usually enforced upon all. This represents a serious infringement of individual freedom, as explained in Chapter Seven.

Most social practices, morality laws, restrictions on individual behavior, and especially practices of raising children and dealing with social stratification, are not essential aspects of belief systems, but are secondarily blended with belief systems. They usually have their origin in ecological conditions that prevailed in the distant past.

For instance, a strict division of labor between husband and wife, including such subsidiaries as dress codes and limits on women's movements outside the home, may have been socially beneficial and even selectively advantageous under once prevailing conditions. But in the twenty-first century, such restrictive rules make no sense, are unnecessarily discriminatory and hold back social development. Unfortunately, changing such rules is a very convoluted process, as these rules are usually thought of as a Deity-decreed part of society.

Some dominant world religions have fared well in the past by urging people to have large families. If this mandate continues in the face of our deteriorating environment and limited resources, the harshness of the ecological feedback will soon become unbearable. We could easily see the human population and its economy crash within the present century. To avoid such a scenario of disaster, religious leaders can no longer afford to believe that this mandate would still be any Creator's design. The future is our responsibility and will be formed by our choices and our ability to live as intelligent, compassionate beings.

Delusions

For centuries, self-styled leaders and prophets have convinced groups of people to follow them in some irrational venture, instigated by a theory or doctrine based merely on such flimsy evidence as a dream, a vision, or cryptic messages in an ancient text of dubious origin. One has to wonder what obsesses people to burn their houses and all their worldly possessions, climb naked up onto a mountaintop, and stand there shivering in the cold, waiting for the world to end at midnight, only to face the next morning with far more serious problems that they had the day before.

An especially puzzling example was the California Heaven's Gate cult, started in 1972 by Marshall Applewhite and his partner, Bonnie Nettles. Based on their interpretation of parts of the Bible, a fascination with UFOs and their personal fantasies, they believed and lured their followers into believing that their bodies were merely temporary containers for their souls, which would soon take flight from their bodies and travel to another part of the universe, designated as heaven. Applewhite claimed to have been the Christ of two thousand years ago, who had arrived again on Earth in the 1920s to lead the Heaven's Gate group into his heaven. In 1997, when the comet, Hale-Bop, became visible from Earth, he concluded that a small object reported by an astronomer near the comet (later found to be an optical distortion) was a space ship coming to take them to their new life.

Some forty followers gathered with him in their communal home and committed mass suicide to separate their souls from their bodies, thereby taking the first step on their journey.

A distressing aspect about the Heaven's Gate case is that the members of the group were generally well educated. These people, and others who get sucked into extreme religious cults, usually desire a shortcut solution to their particular earthly problems. They sacrifice using their intelligence to solve the problems of life in order to follow a charismatic leader, who promises them a problem-solving shortcut, or a way to escape all negativity. These leaders are masterful manipulators and prey on unhappy people. The untenable truth is that words and promises are easy, while actions and life are more difficult. It is a shame that these "avoid reality" beliefs appeal to many, but are never examined with the same scrutiny we demand of scientific theory.[22]

When things become more difficult than we think we can bear, we become more vulnerable to believing delusions—sometimes dangerous delusions. End-of-the world prophesies, utopias, ideal political and religious states--which staunch believers think could be achieved if others, who refuse to accept their delusions were eliminated—fall into this category. Many such delusions have been primary causes of wars— ergo, the Crusades.

A less irrational, but also dangerous belief is that we are in the grip of a predestined course of history, over which we have little control. The original *Utopia*, written in 1516 by Sir Thomas More, was an attempt to describe an ideal society, but the very title means "*no place*" (in Latin) and indicates that More did not actually expect such a state to be realized. Nonetheless, throughout recorded history, utopias have been dreamed up, written about and become the foundation for major political and religious movements—and wars.

In *The Open Society and its Enemies*, Sir Karl Popper traces the destructive utopian theory of Marx through the centuries, via Hegel to Aristotle and eventually to Plato's flawed concept of essentialism.[23] The essence of this utopian idea is that human society inexorably progresses towards some ideal state. Marxism quickly became Leninism and then Stalinism, with disastrous results for humanity, especially when these isms collided with Hitler's Thousand Year Reich, another pie-in-the-sky utopian dream that turned out to be a nightmare.

All of the failed utopian experiments and delusions that promised different kinds of ideal societies have proven that they are unable to solve our problems, stop wars, or increase our resources. It is therefore understandable that many people are leery about the predictions of resource analysts, climatologists and demographers who are telling us to chart a drastically different course, which will move us toward a future of sustainable, peaceful coexistence. But, there is a crucial difference between the promises of unproven theories and these current calls to action.

For the first time in history, our world is being studied by scientists, applying to a massive collection of data the best modeling and analysis methods. So far, many of the predictions made by the scientific community are proving to be accurate within reasonable boundaries. The scientists are also aware of two problems that might unexpectedly steer the direction of our world's unfolding scenario. The first is the potential for sudden positive feedbacks kicking in, such as when we push our current climatic equilibrium beyond its stable range, thereby creating a period of self-accelerating warming to beyond what civilization could survive.

The second would be a totally unexpected scientific or technological invention that could have a major unprecedented effect. Hence, the scientific community is thorough and cautious. It merely warns us that the best data and the most reasonable analyses predict serious problems up ahead that need to be addressed now to avoid the unbearable consequence—a world of deteriorating environments, increasing hardships and even more destructive wars.

Karma

The above consequence can be prevented if we are willing to embrace an ancient philosophy seriously. The concept of "karma" was first introduced in the *Upanishads* of Sanskrit literature. The essence of karma is that you are responsible for the consequences of your actions, and these consequences will determine your future. This translates into the reality that each of us must recognize our part in creating the world's problems and try to solve them with a combination of weighing the relevant information with rational analysis so as to create a desirable future. If we make mistakes, the consequences are ours to live with—or to correct.[24]

If we want to create harmony with and on our planet, we might do well to take a karma-like approach to solving the current problems. We must also accept that we are no longer living in a world where individual decisions result only in limited consequences. In our crowded and highly connected world, we have to plan together and solve our problems together. Creating a world with fewer wars, or hopefully, without war, involves holding a vision for a more cooperative international family.

My (Paddy's) most poignant memories of early childhood center around the air raid drills of World War II. Our family, then consisting of my younger brother, our mother, father and myself, would huddle under a tent, constructed of a blanket draped over four chairs, on the tiny back porch of our home in Sanford, North Carolina. My mother would place a lighted candle in the center of our hideout, and my father would try to explain war. Since my father was the minister of the First Baptist Church, he would begin these little sessions by telling us that God would look after us and keep us safe. My immature five-year-old mind could not comprehend why any God would keep us safe and allow bombs to kill other families and children across the ocean.

Daddy would explain that it wasn't God who killed them, but that people sometimes declared war on other people when they disagreed about how things should be done and could not find a way to live peacefully in the same world. I could not begin to fathom why people would kill each other just because they could not agree on something. I remember looking at my four-year-old brother, aware that we were different, often disagreed, and I had wished he would just disappear at times, but killing him? This would not be possible. I loved him and the other members of our family—well, most of the time.

Throughout the over six decades between World War II and today, my family has grown to include members of three different races and from five different countries, each of which has been involved in countless wars. When we currently get together, we still do not all agree on how things should be done. We argue, laugh and cry together, as members of all families do. We sometimes have to take a time-out from each other to avoid saying or doing something we might regret, but the thought of any member killing another is abhorrent.

Today, my more (hopefully) mature mind is still haunted by the same basic question that I pondered at age five: Why we seem to believe we have a sanctioned right to kill others because we are threatened by differences or cannot compromise to solve problems. The larger any group becomes, the more differences tend to emerge and the more we may have to compromise. If we could hold the true perception that we all belong to one large family of humankind, huddled together under one large sky, on one tiny planet with resources that must be shared in order for us to survive, we might be better able to tap into that place in our minds and hearts that desires to live with less violence—and more compassion. Due to our evolved nature, we hold no illusions that this will be easy. It means stretching our hearts and minds to understand differences, to change many of our perceptions—and to sacrifice to prevent violence, for violence always eventually begets more violence. It requires seeking every possible means of sharing resources without hoarding and depriving others.

Like the *Civil Paths to Peace* report mentioned earlier in this chapter, we must acknowledge that just compromising does not guarantee solutions. In a world of diminishing resources, stretching our hearts and minds, making sacrifices and sharing resources is a must. Ultimately, however, it becomes an ecological issue: if there is not enough to go around, many will suffer and some will die. As soon as possible we must create a sustainable ecology for humankind and the other species of the biosphere. This implies creating a lasting equilibrium between the needs of all of us and what the Earth can provide us in perpetuity. Inevitably, it means decreasing the human footprint—by lowering our population through a much-reduced birthrate, by a planned reduction in per capita consumption and by the application of new technological wizardry. Only then can we have a world without war, in which our descendents will be able to survive. The price may seem high, but not as high as the looming alternative of annihilation.

References:

Q   Wright, Ronald. *A Short History of Progress*. Toronto, Canada: House of Anansi Press, 2004.

1.   Wright, Ronald. Ibid.

2.   Brown, Lester. *Plan 3.0: Mobilizing to Save Civilization*. New York: W.W. Norton & Co, 2008.

3.   Wilson, E. O. *The Future of Life*. New York: Knopf, 2002.

4.   Mech, L. D. *The Arctic Wolf—Ten Years with the Pack*. Stillwater, Minnesota: Voyageur Press, 1997.

5.   Mech, L. D. *The Wolf—The Ecology and Behavior of an Endangered Species*. Minneapolis, Minnesota: University of Minnesota Press, 1981.

6.   Otterbein, K.F. *How War Began*. College Station, Texas: A&M Univ. Press, 2004.

7.   Rees, M. *Our Final Hour: A Scientist's Warning: How terror, error, and environmental disaster threaten humankind's future in this century—on Earth and beyond*. New York: Basic Books, 2003.

8.   Walker, Gabrielle and King, David. *The Hot Topic: What Can We Do About Global Warming and Still Keep the Lights On?* London: Bloomsbury/Harcourt, 2008.

9.   Deffeyes, K. S. *Beyond Oil—The View from Hubbert's Peak*. New York: Hill and Wong, 2005.

10.   Sheeran, Josette, in UN World Report, 2008.

11.   Meadows, D. L. and Meadows, D. H. *The Limits to Growth – Club of Rome Project on the Predicament of Mankind*. New York: Universe Books, 1972.

12.   Kimbrell, A. (Ed.) *Fatal Harvest—The Tragedy of Industrial Agriculture*. Washington, D.C., Island Press. 2002.

13. Madison, James, in Political Observations, 1795.

14. Global Issues, Jan. 2008.

15. Pearse, F. *The Bog Barons—Scandal in the Jungle.* New Scientist 196: 50, 2007.

16. Woodwell, G. M. (Ed.) *Forests in a Full World.* New Haven: Yale University Press, 2001.

17. Fisman, R. and Miguel, E. *Economic Gangsters.* New Jersey: Princeton University Press. 2009.

18. *Report of the Canberra Commission on the Elimination of Nuclear Weapons.* 1997. http://prop.org/2000/canbrp01.htm

19. White, Ralph, Editor. *Psychology and the Prevention of Nuclear War.* 1986.

20. Dyer, Gwynne. *War.* New York: Crown Publishers, 1985. PBS National Television Series. 1986.

21. *Civil Paths to Peace.* Report of the Commonwealth Commission on Respect and Understanding. Commonwealth Secretariat, 2007.

22. Dawkins, Richard. *The God Delusion.* New York: Houghton Mifflin. 2006.

23. Popper, Karl. *The Open Society and its Enemies.* Fifth Edition. Routledge & Kegan, 1966

24. Thurman, Robert. *Inner Revolution.* New York: Riverhead Books, 1998.

CHAPTER

Transforming our Destiny

*If you have a nation of citizens who have risen to that height of
moral civilization that they will not declare war or carry arms,
for they have not so much madness left in their brains, you have a
nation of true, great and able people.*
~ Ralph Waldo Emerson

Transforming the destiny of the world first requires that we fully understand the reality of how we have created a non-sustainable planet and what we must change in order to correct this fatally destructive trend, or our descendents will not likely survive another century. In order for them to inherit a sustainable world, we must look within ourselves to discover what motivates us as individuals. As emphasized at the conclusion of Chapter One, if we want to understand the world, we need to first understand our individual selves. Our prime motivation is to have our existence acknowledged—to experience our aliveness through validation and acceptance by other humans. This ideally happens when we belong to a community of fellow humans and feel accepted by at least some of them. Our quests for love and for safety are powerful drives. When love is not forthcoming, or when the objects of our love are threatened, we feel fear and become either passive or aggressive. These conditions shift our motivation to separating from or destroying the perceived cause of our fear before it destroys us. These primary emotional drives to create both love and war are deeply ingrained in our nature and remain with us from the cradle to the grave.

Most of us make choices which we believe will be life-enhancing for us and for those we love. Usually, this includes members of our families and others to whom we feel some allegiance. In order to transform our destiny, now is the time at which we must make a major perceptual shift—to expand the group to which we belong and owe allegiance to include all humanity and future generations.

We need to accept—at the deepest level of our being—that each of us is a member of the global unit known as the Earth's biosphere, on which we depend for survival. We are also rapidly becoming one international community. Our species has spent most of its civilized history devastating our planetary home through greed and wars, without comprehending the damage being wrought. Because we have created a critical situation that threatens our survival, each of us has an ethical and moral responsibility to change our perceptions and behaviors in order to form a healing relationship with Earth and her residents. According to cultural historian Thomas Berry, in his recent book, *The Great Work: Our Way into the Future*, this is the "great work" we must accomplish if our species is to survive beyond this century.[1]

The first step of this work begins inside our individual souls, with the question: Am I willing to make the necessary changes and sacrifices that will ensure a future for the planet and my descendents, or will I continue to hide behind the myths that allow me to write off the future with an attitude of *après nous, le déluge* (after us, the flood)? This comment is attributed to Madame de Pompadour, mistress of French King Louis XV, expressing her total disregard for what would happen to the world after her death. It is easy to label her attitude as selfish and offensive, but many of our own shortsighted perceptions and behaviors are equally selfish—and may well become deadly. It is late, but we hope not too late, to prove that our love *For the Beauty of the Earth*, which many of us learned to sing in church or school, expresses sincere gratitude and is strong enough for us to seek long-term harmony on our planet rather than short-term gratification.

The most valuable legacy we can leave for future generations is to begin to heal our devastating exploitation of Earth so that they will be able to continue a mutually beneficial relationship with the planet.

For more than forty years I (Dolf) have written a series of letters, which essentially comprise my diary. These letters were usually addressed to my parents, and following my father's death, to my mother. After her death I did not write for several months. A half-written letter, started on the day she died, still sits in my computer.

After a period of time, I began to write letters to my sons and will share a portion of one:

I recall a nice, sunny day a long time ago when I was a young teenager. The family went on a cycling trip from Kampen in the Netherlands, where we lived, pedalling for quite a distance along canals and through fields and pastures. It was an entire family outing, including Father, Mother, my brother, my younger sister, who was handicapped and in need of parental assistance, and myself. I do not recall with any precision exactly where we went, but I do recall the feeling of summer wind in my face, of flowers and green grass, of sunlight on water and of the sound of gravel under my bicycle tires. I felt my oats and sensed the future ahead of me.

On our way back home, my brother and I raced ahead and were home long before the others. I remember my parents' anger at us for racing ahead. We got a serious dressing down for violating the principle of "samen uit, samen thuis," which is Dutch for "when you start a journey together, you end it together." I felt at the time that I had been thoughtless and selfish, but I also felt that all members of a group should enjoy themselves in their own way. I did, however appreciate then,

259

and still do today, the symbolic meaning of that situation.

As you boys were growing up, we went on a number of trips together, and the few times we had to, or chose to, split up (as we did on the slopes of Mount Wilhelm or for that Oxford summer) I felt that same conflict situation and guilt of racing ahead. Family life is like a long trip, but sooner or later the younger ones must race ahead—and sooner or later, the older ones don't come home.

We are never sure how our choices may affect others, but frequently our selfish needs are in conflict with our selflessness to consider others. It can be difficult to balance these needs. We can err unintentionally, as Dolf and his brother did. Slight offenses, like theirs, are usually easily forgiven. But offenses become more damaging when we are unwilling to face how others could be negatively affected—in the present and in the future by our actions.

Whether we will have the foresight, courage and love in our hearts to sacrifice now for our descendents so that they might have a sustainable future is up to each of us. The day will come when those of us reading these words will not come home, but we will be remembered by the kind of world we left behind.

Why Change Now?

In the previous chapter we laid out the reasons for making important changes as soon as possible. It comes down to this: the processes we have used to sustain our population and our civilization are no longer sustainable. Many are already showing signs of deterioration, while simultaneously producing accelerating levels of deleterious by-products. These processes can be categorized as harvesting nature (e.g. fisheries, forestry, clearing land), agriculture, industry and transportation, extracting non-renewable resources (fossil energy, minerals) and carrying out military operations. As each of these activities has intensified over the past century, they have begun to affect one another negatively. Harvesting nature is beginning to destabilize the natural equilibrium of the Earth's ecosystems, while industry is causing global warming and pollution, which reduces nature's ability to recover from direct damage. Agriculture is negatively affected by climate change and by industrial pollution, spilling its negative side effects, such as soil exhaustion and the agro-chemicals, onto nature. Our perceived need to spend billions on the military not only gobbles up resources that could be used to nurture life, but also increases environmental destruction.

Continuing to promote further growth of the population and of the world economy is sheer folly as it excessively enlarges our damaging footprint on the planet. This results in expansion of the very processes which are already pushing against their ceilings. This complex, inter-active threat to our future by itself will cause a major collapse within the next one hundred years. Given our long history of choosing to fight rather than starve, it is almost certain that major wars will break out over the control of diminishing resources. In a world with nuclear weapons, this could be suicidal. These harsh realities present an abbreviated, but potent, argument for the need to make major changes now.

Dealing with Human Nature

The most serious detriment to solving these problems is that we still possess traits that were perfectly adaptive in more primitive times, but are woefully inadequate for facing our present challenges. One of these traits is our focus on quick fixes, while ignoring the long-term negative consequences. Since primitive man could neither predict nor affect long-term events (beyond the annual cycle or the recognition of multi-year dry and wet periods) it is not surprising that the ability to recognize and manipulate longer-term trends has not evolved strongly in our species. Related to this shortcoming of our nature is our reluc-tance to accept the need for change as long as we perceive that things are going well enough. It tends to make most of us angry when others try to convince us of future impending danger, or when the myths we prefer to believe are threatened by the truth we had rather deny—such as the fact that we have blindly participated in creating a fast slide toward catastro-phe. Rather than deal with the consequences of our actions, it is easier to rationalize that since we have survived ghastly wars, natural disasters and other destructive events, future generations will be able to overcome whatever obstacles threaten their survival. This is the fairytale of our modern age. It is time to accept that our present major problems are the result of continuing to depend upon past solutions—no longer appropriate for future survival.

Throughout this book, we have built a case for accepting our fairly-hardwired nature, which forms the genetically determined base for most of our individual behaviors and personalities under a wide variety of environments. We have introduced the different forms of natural selection, which over the eons have shaped human nature with its high degree of individual variation, its apparent contradictions and mal-adaptations. Natural selection shapes us for survival, but it takes considerably longer than we now have for evolution to change us enough to survive the present crises. What we must recognize is that we each have to live with the version of human nature we carry in our genes.

We have used the term "evil" sparingly, because it is a complicated concept that can often causes problems in the ways it is used. In Chapter One, we stated that good and evil are not concepts that apply to the material world, but that is not entirely true. When we recognize good or evil behavior in social animals, including humans, we can surmise that, in all probability, there is a DNA based genetic component. This implies that good and evil are aspects of human nature and can be influenced by natural selection, especially by socio-culturally driven selection. In our opinion, real evil consists of recognized destructive anti-social behavior within a cohesive, sovereign community. An overly despotic chimpanzee alpha male, an adult human who rapes a child or a serial killer are usually considered evil and certainly have a deep and destructive evil aspect to their nature. A complicating factor is that these same animals and humans are also capable of showing kindness in a different circumstance.

Another complication with the term is that some actions can be experienced as evil in one society, but considered perfectly normal in a society with different values. Circumcising a young girl is considered evil in Western society, but not in Somalia. We tend to denigrate our enemies by calling them evil, who in turn see us as evil.

Enemy acts can be barbaric, vile, and beyond the pale, but not really evil, because the enemy lies beyond the moral/ethical domain of our community. We fight our enemies with everything we can throw at them, even with actions that within our own community we would consider evil. Remember the different uses of steel weapons by the Murle people of Southern Sudan, or the terror bombing by the "evil" Germans as compared to our "strategic" bombing of German Cities.

This deeply ingrained double standard in our moral instinct is also a part of human nature, which will be a serious impediment when we try to embrace all of humanity into a global community. Even what we consider the first step towards creating a world community without war, i.e. creating a sustainable human ecology, will be very difficult as many nations will distrust others. One of the main stumbling blocks in trying to get an effective treaty on global warming at the Copenhagen conference was China's refusal to get international surveillance of their carbon emissions, as they considered that a violation of their sovereignty. In response, the United States immediately assumed that China would cheat. These gut reactions are based partly on our hard-wired distrust of strangers and can be reinforced through past experiences. To overcome these reactions and emotions in order to establish a rational and safe way of charting our global future presents a huge stumbling block.

Since the latest major evolutionary development has been our superior neo-cortex, which gives us the ability to think rationally and to use our higher consciousness, we are no longer stuck in the primitive competitive model of our ape ancestors with little chance of consciously changing their lives. This higher structure of our brain automatically builds upon and integrates the lower structures, while providing the network for us to strengthen the traits that will be advantageous to our survival. As the reality of our present circumstances dictates that changes in our perceptions and behaviors are necessary for future survival, evolution has given us the equipment to make intelligent choices that will support the well-being of our species and of our planetary home, on which our well-being is dependent. However, when we deny reality, or continue to believe we do not need to change, we are not using the equipment provided by evolution to help us survive.[2]

When resource shortages loom and xenophobia rules our psyches, we function as *Homo hostiles*. When we grasp that we all share human status and one planetary home, realize we must stop the destructive waste of resources and focus on developing more economical use of what is available, we can function as *Homo amicus*, the kindly human, who holds the potential for creating a more compassionate, peaceful and sustainable world.[3]

We will only survive as a species if we accept our relatedness and cooperate to create an ecology that can support the global population, while we simultaneously reinforce the traits of empathy, forgiveness, cooperation and compassion. Unless we each commit to doing everything we possibly can to reach this goal, we will revert to wars over limited resources, stuck in the past without a viable future.

Our Most Difficult Battles

We stand at a critical crossroads. We can choose to live as world citizens and begin to overcome our damaging habits, or we can choose to carry on as in the past, ignoring the rapidly accumulating warning signs. We each must choose which wolf to feed— moment by moment. No law can dictate this choice; it can only be made within each of our minds and hearts. The tally of our individual choices will transform our destiny—or destroy us. It is overly idealistic to think that each of us will be able to live every moment of our lives as a kindly, loving person. Most of us will continue to make some colossal mistakes, to tell a few lies, to act on short-term and selfish motives, to feel intensely negative at times, to spread our fair share of misery onto others and to lose sight of positive goals. Our progression toward a goal, either as an individual or collectively, is rarely composed of all positive steps. Our negative emotions, thoughts and passions remain a powerful part of our nature. Our hope is that we each will use our rationality and higher consciousness to engage in constant battles with the negative aspects of our nature to prevent us from causing more damage. According to Marc Hauser, director of Harvard's Cognitive Evolution Laboratory, in his recent book, *Moral Minds*, we each have an evolved moral instinct that develops within our higher consciousness as we interact over time within our communities. This instinct gives us an innate sense of right and wrong. It is often referred to as our conscience, or the "still, small voice within," and is a major player when we have to make choices. If we check in with our moral mind before choosing a course of action, we will raise our chances of making fewer mistakes.[4] Unfortunately, this very recent research has not as yet addressed the remarkable and disturbing difference between the within-group and the between-group aspects of our moral instinct.

We are not all created equal, nor do we all have equal opportunities, but each of us possesses the brainpower and a higher consciousness that recognizes our conflicting emotions and inappropriate urges and can deduce the most-likely consequences of our behaviors. We each participate in creating some portion of our small corner of the world, and some have far-reaching influence. Whether we consciously choose to influence others or not, most of what we do and say holds the potential to affect others, who in turn affect others—and so on and so on. Due to the advanced technological changes in our methods of communication, we live in a more tightly woven web of connections than ever before in history, which means that we each can be affected within minutes by things happening, or words spoken, almost anywhere in the world. This is equally wonderful, worrisome—and risky. The wonderful part is that we each possess a degree of personal power and ability to make things happen. The worrisome aspect is that we never know how others might react. The risky part is that we can never control all of the consequences of what happens—to ourselves or to others. It takes one human being to dream of personal glory through waging war, but it takes thousands of us to empower such a person. From that perspective, each of us contributes to waging war—or to promoting peace.

Regardless of how insignificant any of us may feel as an individual, we each have some influence upon other members of this global unit. In his book, *The Politics of History*, political scientist Howard Zinn warns us that none of us can afford to be an innocent bystander. He writes:

> *We don't have to engage in grand, heroic actions to participate in the process of change. Small acts, when multiplied by millions of individuals, can transform the world.*[5]

Three Cups of Tea

Greg Mortenson's memoir, *Three Cups of Tea*, illustrates how one person committed to fulfilling a promise made over a cup of tea brought education and improved life to an impoverished sector of the world—one school at a time. Mortenson was not in any elected leadership position, nor a hero of any kind. He was a regular guy who set out on an ordinary personal journey that evolved, step by step, into an extraordinary experience for thousands.

In the northern border areas of Pakistan, there is a custom known as "three cups of tea." When interacting with a new person for business or social purposes, you drink tea together. During the first cup, you are drinking with a stranger. If offered a second cup, you drink with a friend. If presented a third cup, you are considered as "family."

Due to deteriorating weather on a September day in 1993, Greg Mortenson failed to reach the summit of K2 by only six hundred meters. He had planned to place a special memorial at the summit to honor his twenty-three-year-old sister, who had died the year before from epilepsy. Overwhelmed with disappointment and fatigue, his eyes filled with tears, he lost sight of his climbing team, who believed he was close behind them. Eight days later, having endured extreme circumstances of hunger, thirst, cold and fatigue, which would have certainly ended the lives of most of us, he stumbled into the village of Korphe, high in the Karakoram Mountains. There he was greeted by a group of children, who had never seen a foreigner. Awed, the children led him to the home of Haji Ali, chief of their village, who offered Mortenson his first cup of butter tea. And then another—and another.

After the third cup, he fell asleep on this kind man's hearth. It was the first night he had slept indoors in weeks. Three weeks later, after regaining his strength and drinking many cups of tea with these people, to whom he owed his life, he realized there was a much more meaningful memorial he could make to his sister. He promised Haji Ali that one day he would return to the village and build a school, which even girls could attend. With this promise, Mortenson's life changed more profoundly than he ever could have imagined. Over the next decade, during which he encountered every problem imaginable —from raising money to trucking in materials over nearly impassible mountain roads— he became responsible for the building of fifty-five schools. Since then, thousands of children, mostly girls, have been educated, and thus able to live healthier lives.

It has been established that education is not only the key to enhanced opportunities for young women, but also to reducing unwanted pregnancies. In this part of the world, where westerners have never been understood, and where few outsiders have any understanding of life in these isolated mountain villages, this one man has created an awareness that has changed untold numbers of perceptions—primarily, that females can be educated and have meaningful lives and that this education improves living conditions for the entire village. Mortenson has freely given the kind of love that is the basis for building a better world—a world without war.[6]

Love in War

War, like love, becomes so perverted in our minds—so filled with myths we wish were true. In *War is a Force that Gives Us Meaning*, New York Times staff writer and fifteen-year foreign correspondent Chris Hedges writes that both love and war initially hold the power to fill our deepest spiritual needs: to share a sense of meaning and purpose, an opportunity to choose self-sacrifice over security and to resolve the internal tug-of-war between the Freudian concepts of Eros (the urge for self preservation) and Thanatos (the urge for destruction) deeply embedded in our nature. The closer we come to the edge of any life-death resolution, the more passionate we tend to feel about life. That edge is ever present on the battlefield where love of country and of comrades is up against hatred of the enemy and the fear of death. When caught in this intensely passionate conflict, we also remain highly motivated to be accepted and validated by those who hold power over us. In war, these dynamics often turn ordinary people into violent killers.[7]

This truth is explored in Slavenka Drakulic's *They Would Never Hurt a Fly*. Once a citizen of Yugoslavia, which is now divided into several new states by a bloody war during the early 1990s, she writes of sitting in the Tribunal Building in The Hague, where several of those accused of murder and torture of prisoners during this war are being tried. There are five defendants, whom she describes as looking quite ordinary—no horns or tails. The one who most intrigues her is Goran Jelisic, a thirty-six-year-old man who looks as if you could trust him to watch your children. He had been a child of the late sixties, considered the first normal generation after the fall of communism. He had become a successful farm mechanic, whose favorite past time was fishing. He often gave the fish he caught to friends and neighbors.

The one thing that seemed to bother Jelisic, according to witnesses on his behalf, was that he never made enough money to have the things he wanted. This desire drove him to start forging checks, for which he was sentenced to several months in prison. While in prison, he learned to work a computer and was considered a good example for others. He was released early when a strategy was implemented to provide volunteers for the war through releasing prisoners. He volunteered to be a military policeman.

Over the next two years, Jelisic shot and killed many prisoners at close range in the Luka camp, near where he was stationed. It was said that he would enter a room of prisoners, announce himself as "the second Adolf Hitler," and prisoners would tremble at the sound of his voice. He would point out victims at random and, without making any accusation, shoot them in cold blood. If there were other policemen or female guests watching him, he would often torture the prisoner first, by cutting off body parts and exhibiting them. He openly expressed his enjoyment of torturing and killing. He bragged about having killed over a hundred prisoners in eighteen days, during May, 1992.

The judge at his trial stated to the court that he had never before heard witnesses for anyone testify how good a man was prior to the war, followed by testimony of others who observed the same man's behavior during the war and described him as a cruel and heartless executioner. Jelisic Goran was one of three, from a group of fifty-five men on trial, who admitted that he was guilty as accused. He showed no remorse. It was pointed out by his lawyers that he had never killed before the war, and to their knowledge had never killed since. The only explanation offered by the court is that this man was transformed by the circumstances of war. Or is it possible that he was highly motivated to gain validation and respect from those with whom he identified? When he was commanded to kill, he killed, and was awarded by the clout he gained through being good at what he was doing. He was sentenced to forty years in prison.[8]

The nature of war uses our need to be accepted and validated by the groups with which we identify, paired with our xenophobia, to see other human beings (the enemy) as objects to be slaughtered. It forces us to extinguish our ability to love in order to survive. For this very reason, many have referred to war as an epidemic of madness. War worships power and allows the powerful to become victims to their power. It feeds the greed of leaders, who reward the heroes, who kill. It intoxicates a populace, who root for carnage and become blinded to our shared human status. War is the triumph of a primitive part of human nature, which we must no longer tolerate as a driving force in how we manage the world.

Can Love Help Us End War?

Psychological philosopher, Victor Frankl, who survived Auschwitz, writes in *Man's Search for Meaning* that our ability to love is our only salvation from our primal and savage impulses, which dominate in war. He admits, however, that it is extremely difficult for love to survive unless we are in an environment where it can be nurtured. And that it is impossible for children to understand love when they are surrounded by the violence and hate of war, without a loving family to protect them from the cruelty of the war around them.[9]

This is poignantly driven home in the true story of Ishmael Beah and the over three hundred thousand other child soldiers with a similar history. Ishmael was raised in a loving extended family in the village of Mogbwemo in Sierra Leone. He went to school, loved movies and performing rap music. At the age of twelve, he was hiking to a nearby village with his brother and a friend to participate in a talent show when they heard from people running past them that their village had been attacked by rebels and most of its inhabitants had been brutally murdered. He, his brother and their friends became orphans, hiding and running for their lives on paths, strewn with dead and mutilated bodies, through Sierra Leone. When they arrived at an army base in the southeastern region of the country, they thought they were finally safe.

However, an unimaginable nightmare had just begun for each of them. The army conscripted all of the boys and taught them to kill. Ishmael watched as his two closest friends were blown apart by grenades. He learned to hate. He was given drugs that numbed his senses, but revved him up to the point that he could not sleep for days, during which he could kill. The drugs dulled his brain, but kept his body marching until, according to his personal account, he writes that killing had become as easy as drinking water. He reports that his squad had become his family, his gun his protector and his heart had frozen. After months of this no-life, his childhood was over.

Later, a miracle intervened: a truck of UNICEF workers came to their camp and demanded that the army give up the younger boys. Ishmael and others of his group were given to the four men in the truck, but the boys were terrified without their guns, which had been taken away by their lieutenant. After driving for hours, they arrived at a UNICEF center, where they met other boy soldiers, who had belonged to the rebel army—their enemies. Just as the boys in the Robbers Cave experiment, described in Chapter Eight, when they discovered another group on their campground, these two groups of boy soldiers instantly felt animosity toward each other. Because they had been brainwashed and trained to kill, they immediately instigated a dangerous battle, during which several were killed. The biggest difference between the Robber's Cave boys and the Sierra Leone ones was their perception of what should be done. The campers had not been trained to kill, the boy soldiers had. This is the training of war. No one walks off a battlefield without wounds.

After months of rehabilitation, an uncle of Ishmael's came to take him back to Sierra Leone, where he was cared for by the loving family of his relatives. He writes that he continued to have nightmares of killing and would often see gushing blood rather than water coming from a faucet during the day. But, over time with these loving people, he learned to accept that he could be loved and love in return. One of the major impediments to accepting love was the guilt he was now feeling as his perception changed of what he had done as a soldier. He was often told that what he had to do to survive in the army was not his fault, but this phrase haunted him. He knew in his deepest soul that he had to let go of love to be able to survive war, however, being loved again, he could not forgive himself for what he had done to other human beings.

Years later, Ishmael was invited to speak of his experiences to a United Nations assembly in New York. An American woman was so deeply moved by his story that she adopted him, which again changed his life profoundly. He completed high school and then graduated from Oberlin College in 2004. In 2007, his book, *A Long Way Gone: Memoirs of a Boy Soldier*, was published, to help the world understand how war robs children, and often adults, of their ability to love.[10]

War is a thief of love, but can love help us overcome war? Yes, but only if we chose to give up our greed and the need for dominance over others in a desire to understand them with empathy and compassion. Compassion is the only force that can give us the power to resist within our own nature what we must resist to end the atrocities of war.

Hope from the Animal World

We have learned that most of our progenitors fought to kill when threatened by resource shortages or from fear of losing what was valued. Never the less, the bonobo, only recognized as a species for less than a century, is presently being referred to by primate specialists as the "make-love-not-war primate," because this species uses a wide array of sexual activities to resolve most of their conflicts and competitive situations.

Like their close chimpanzee relatives, they move around the forest in small groups that form parts of sovereign communities, from which the juvenile females emigrate to be absorbed by neighboring communities, while the males stay behind, but bonobos have evolved a remarkably different social structure. The bonobo females are more strongly bonded than those in chimpanzee communities, and female bonobos dominate the males in most interactions. Males never gang up on a female, but females have been observed ganging up on a male to convince him to be more cooperative. Thus far, they have demonstrated less chest-thumping and aggressive behaviors, while spending a great deal of time engaged in kissing and copulating—both heterosexually and homosexually. They appear to demonstrate some degree of sensitivity to all members of their groups.

Takeshi Furuichi, a leading bonobo researcher, says that bonobos, in the wild as well as in captivity, seem to be enjoying life and make efforts to keep peace by showing frequent affection toward each other. After observing and studying the bonobo, primatologist Franz de Waal, wrote that while the chimpanzee resolves sexual needs with power, the bonobo resolves power issues with sex. In his book, *Bonobo: The Forgotten Ape*, he writes:

> *Who could have imagined a close relative of ours in which female alliances intimidate males, sexual behavior is as rich as ours, different groups do not fight, but mingle, mothers take on a central role, and the greatest intellectual achievement is not tool use, but sensitivity to others?*[11]

Taking the evidence available to date of bonobo behavior, de Waal admits that we do not know how this sexually expressive animal manages to avoid over-use of its resources due to excessive population growth and thus avoids fighting with neighboring communities—if indeed they do manage these, which remains the critical question. Some observers have reported noticing healed injuries on males that could be the result of inter-community fighting, but could also be the result of other kinds of physical trauma. A possible explanation for their population control is that they might be decimated for meat by the local human population. Another more appealing explanation is that when their population becomes high enough to affect resource density, the bonobo might have a behavioral or physiological response, similar to the Arctic wolf, which lowers the birth rate.[12]

What we learn from this is that a social species living peacefully in a stable relationship with its resources, whether ape, whale, wolf or bonobo, has to have feedback from its ecology into its demography. In other words, either mortality increases or births decrease when the population reaches a level beyond which it would start to overexploit essential resources. Can we create such a relationship with Mother Nature for ourselves? Only if we give up an important freedom we have been taking for granted--to have as many children as we want.

The Challenge of Everyman

In fifteenth century Europe, as society was emerging from the Middle Ages, religious and state authorities were worried about the growing desire for personal wealth, the reduction in church attendance, widespread immorality and violent crime, especially in the rapidly growing cities. These same concerns are equally relevant in this twenty-first century.

Considerable pressure was applied to the populace through laws and law enforcement, sermons from the pulpits and through popular morality plays. In retrospect, we can see these plays as an active form of socially driven selection. They created a social atmosphere that praised those who followed the examples set by the plays' moral characters, while censuring those who did not.

One of the best-known morality plays of that era was *Everyman*. In this short work, a typically complacent, sinful person called Everyman is approached by Death, who announces that the end of his life on Earth is nigh and that God has summoned him to heaven to make his reckoning. Poor Everyman, knowing he has caused a great deal of damage during his reckless life, is desperate. He tries to bribe Death to spare him, but to no avail. He next approaches those whom he has considered to be his friends: Fellowship, Kindred, Cousin and Goods. But when they discover the destination of Everyman's journey, each finds a reason to abandon him, proving themselves no more than fair-weather friends.

Finally, Everyman approaches Good Deeds, who hesitates at first because Everyman had ignored her in the past. But later, she agrees to accompany him after he agrees to pay a visit to Confession. He confesses his sins and repents, while beating himself with a scourge. Virtuous friends, Beauty, Knowledge and Discretion support his repentance, but decide they should not accompany him. Only Good Deeds goes with Everyman to his grave. His self-inflicted penance allows God to forgive him, and he ascends into heaven with Good Deeds. The play ends with a narrator explaining the message of the play to those who might have missed the point—that we can take with us from this world nothing that we have received, only what we have given.[13]

For most of us in the twenty-first century, an important aspect of looking back on one's life when death seems imminent has to be an evaluation not only of what good deeds we may have done for our contemporaries, but also of what we leave behind for future generations.

It is unlikely that moralists of the fifteenth century were at all concerned about the future of the physical world in which the people of succeeding centuries would find themselves. We must be concerned because we are aware of the alarming rate of deterioration of our planet, which is mostly the result of our larger population living higher on the hog. Trying to integrate into human nature a moral code to identify the sins and good deeds regarding our relation with future generations demands careful consideration and is a challenge for each of us.

The essence of this process is that we define the targets intellectually, based on the best scientific knowledge and methodology, and then must embrace the moral obligation to adjust our lives accordingly so as to reach the ultimate goal. Our political and religious leaders are the ones who will have to take the initiative to reshape the political and economic structures of society, but they will only do so when the vast majority of us demand it of them.

Taking Action Today

In 1967, my wife and I (Dolf) bought an abandoned farm on approximately a hundred acres of hardscrabble land in Eastern Ontario. On some parts of the land, old, partly decayed stumps of large pine trees hovered some fifteen inches above bare fissured limestone, held in position by the remnants of the trees' roots that had found a toe-hold in the deep cracks in the rock and a way down toward essential water and nutrients. As the early nineteenth century owners of the land cut the forest that had been there since shortly after the glaciers receded some twelve thousand years ago, the process of erosion started to reverse the soil-building process that had been the rule over twelve millennia. It took the men of the nineteenth century only a year or two of backbreaking labor to clear the land; but within one century, erosion had done the inevitable. Decaying stumps and piles of rocks remained as a memorial to the sweat and tears of these settlers, whose descendants abandoned the land for a better life elsewhere.

With hindsight and advancement in scientific knowledge over two centuries, it was easy for us to see why these settlers had failed on this land in this climate. By cutting the forest and burning most of the biomass they removed, they contributed to the first signs of global warming. We could not blame Mr. Ferguson, who received the land we now own as a settler's grant in 1803, nor could we blame the authorities for believing that such land could be used for farming. We, however, could see the resulting tragedy, and we restated the golden rule in a new way: *Do for your descendants what you wish your ancestors had done for you.*

We decided to re-establish a forest. After gaining the best available scientific knowledge on what kind of trees would be most likely to succeed on shallow, stony soil, and with the help of a government grant, we planted thirty thousand three-year old seedlings. To our surprise, many of the local farmers and villagers responded critically. There was some resentment, especially among the older men, as they remembered stories from their grandparents of how the settlers had suffered while clearing the land, how they had nearly starved before they got their first crops off the new land. But even more significant was their warning to us that we would never see those trees as the giants in a mature forest. All we could do was quote our golden rule and hope that as our trees grew, the wisdom of our planting would become apparent.

Today, in 2010, some forty years later, we do have a forest, though not a mature one. It is a growing forest that fixes some fifteen tons of carbon each year, which is considerably more than we send into the atmosphere with our one car, the heating of our house and our occasional flights to faraway places. Our neighbors now appreciate the forest and its wildlife. Of greater importance to us is that our progeny will be able to see and enjoy it as a mature forest, large and green enough to be visible from space. Not everyone has the opportunity to plant trees on a hundred-acre patch of land, nor build schools in Pakistan, but each of us can make choices to help nurture our planet.

Coming to Terms

It is time to come to terms with our beliefs about war and peace, with love and enmity, with life and death. The bottom line question for each of us becomes what do we really want and what are we willing to pay to obtain it? War comes at a high price—not only in financial terms, but in terms of life, death and untold suffering. Peace may come at a price as well and each one of us has personal choices to make regarding what we are willing to contribute toward creating it. The fact that we have survived at all attests to our ability to change what must be changed to create new opportunities for survival. If there is any grain of truth in Tolstoy's belief that lustful love and war are the most passionate experiences in life, we will continue to fall "madly" in love, but we need to expand love to empathy and compassion for all others. We may continue to have a fleeting impulse to wage war when we feel threatened, but must consider all peaceful means of negotiating and compromising to resolve problems. If there is truth in Whittier's quote that making peace may require more of us than does waging war, we must come to terms with what must be done and the sacrifices we must make for peace.

According to Sam Keen in *Faces of the Enemy*, wars are primarily caused by differing interests of sovereign nations and not by our inability to communicate or understand each other, when we are willing to make the effort. Greg Mortenson's experience gives credence to this idea. He held a vision of what could be done and refused, against all odds, to give it up. If we were willing to unite to create and hold a vision of what could be done to transform our destiny and heal our planet, we could change our systems of nationalistic anarchy, which lead to violent conflicts and wars. We could develop and support a global political body with the power to adjudicate national conflicts. Creating the League of Nations and the United Nations were steps in this direction, but the most powerful nations in these organizations have not been willing to limit their sovereign right to use unilateral violence or willing to limit the control of their empires. The more an empire seeks to extend power and control, the greater the chances for chaos and rebellion become.

A fascinating aspect of this phenomenon is that it also holds for us as individuals. The tighter we hold control over another person, the more apt we are to lose control; and crimes of passion are often the result. The great philosopher, Hegel, wrote that in the master-slave relationship, the master is as bound as the slave. Slavery, the act of owning other human beings is de-humanizing to the ultimate degree for both slave and master.[14] Trying to maintain the illusion of total control is exhausting—for individuals and for nations. There was a day, not that long ago, when only a precious few believed that slavery could be abolished, however, today it is a crime to own slaves. This being true, it should be equally true that we can abolish war—and hope the day will come when it will be a crime to wage war.[15]

Former Soviet Union leader Mikhail Gorbachev recently gave a speech, which I (Paddy) was privileged to hear. He declared that our world suffers from the lack of a global government and global leadership. "Global leaders," he stated, "must be able to govern, based on a broad analysis and understanding of what is really happening and not be fretted by dogma and ideology, if we are ever to have peace in this world."

He went on to explain that security can only be collective and that there will be no security until the peoples of the world demand that leaders address and do something about security, poverty and the environment. In expressing his concern about the proliferation of nuclear weapons, he explained that these weapons make it easy for the larger powers to protect their remaining resources, but at what price?[16]

We each need to consider what price we are willing to pay for peace, which requires that we give up the madness of power over others, end war and ban nuclear weapons. It requires that we each commit to making the changes we can make to healing our minds, our hearts— and our planet. The United States has recently (2008) elected President Obama, who is demonstrating his commitment to resolving conflicts through peaceful means and is actively encouraging people the world over to develop more understanding and compassion toward all others. He urges each of us to believe in our personal ability to be able to change what must change so that we can survive together on our planet.

A Microcosm of the Macrocosm

A few years ago, my husband and I (Paddy) were in Mexico City, preparing to climb Popocatapetl, an eighteen-thousand-foot volcano, located several miles outside of the city. In order to reach the bus terminal where a bus would take us to the small town at the base of the volcano, we had to catch a subway in one of the busiest and most crowded terminals in the city at rush hour.

On the subway, we were aware that we were the only "light skins," packed like two hundred sardines into a can made to contain one-third that number. There were no seats; everyone stood using one arm to hold onto railings above our heads. Tim and I each had our other arm clutched around our backpacks, which were squeezed between us by the pressure of those around us. I'm ashamed to admit that I felt fearful and vulnerable, knowing how "different" we must appear to those around us. Within seconds, I felt hands all over my body—rubbing, poking and one slipping around my waist trying to unzip my jeans. I then felt the depths of a deep primordial fear, not specifically from being alienated by skin color or culture, for I knew the same thing could happen anywhere when this many people were crushed together, but mostly from not knowing how to safely stop what was happening to me. I was in my fight/flight reaction, but there was no space for either. I solicited help by saying to Tim, "There are hands all over me. I'm terrified."

He immediately jabbed his elbow as hard as possible, within the limited space, into the group around us. We heard a groan, and several hands dropped away from my body, but we did not see a change of expression on any of the opaque faces. Tim then suggested that we turn facing each other, so that our fronts were flush against each other. After that move, I felt connected and safer in having a strong ally.

Finally, the ride was over and we were relieved to be in the bus station with our belongings, but our troubles were not over. Through my feeble attempt to speak Spanish, we purchased tickets and headed for the bus. We were overjoyed that our bus had two front seats available. Settled in and beginning to consult our guidebook for a place to stay in the village, we were startled when a very angry woman began to shake her hand in our faces, screaming that she purchased those seats. Others on the bus began to laugh and call us "gringos stupidos." We were bewildered until the angry woman did her best to explain that we had the cheapest tickets and should sit on the back row of the bus, which already appeared crowded. The people sitting there understood and slipped closer together, leaving space for us. My fear subsided at this act of kindness, and once we acknowledged that we were content in our proper places, those around us begin to smile at us.

About an hour into the ride, someone yelled, and the bus came to a halt in the middle of nowhere. The bus was on fire from overheated brakes. Everyone began to stumble off the bus to gather rushes and branches from fields and trees to thrash out the flames. We worked along with the other passengers, which elicited a few smiles in our direction and gained us some acceptance as members of the group. After about thirty minutes, the fire was out, but the driver decided the brakes were not safe for us to continue. There were no cell phones in the hinterlands of Mexico at that time, so there was nothing to do except wait for other buses to rumble by, try to flag them down and find a seat on any that had space. Tim and I knew by the time we were placed on a bus that it would be too late to find a room for the rest of that night.

After we arrived at the closed bus station in the village of our destination, everything else was also closed. As we settled on the ground, leaning against the wall of the station to wait for morning, the same woman who had been so angry at us for occupying her reserved seats on the bus, approached us. Smiling and holding out her hand, she said, "Mi casa es su casa." (My house is your house.)

We followed her a few blocks down the street to her casa, composed of one large room with a bed, where a man was snoring away, a small cot with a young boy asleep on it, and a shabby sofa. She expertly rolled the man out of his bed and offered it to us. The man, whom she identified as her spouse, was startled and upset, but mumbling to his wife, he fell onto the shabby sofa and closed his eyes. She climbed onto the cot beside the boy, as she motioned for us to get into the larger bed.

After what seemed less than ten minutes of sleep-time, we were awakened by movement and the smell of fresh bread. The boy, her grandson, was standing close to where we were sleeping, holding warm bread, and smiling at us. His grandmother had a small table set for breakfast on an open patio, connected to the outside back wall of the house, and invited us to sit down. During breakfast, she explained that her brother had a cab and would drive us up to the base of the mountain. We offered to pay her, but she refused to accept the money, saying we were the only Americans who had ever been in her house and she was honored to help us.

For me, this experience has become a microcosm depicting the macrocosm of our entire known world and its peoples. As global citizens, we are going to have to deal with too many people in too small a space until we are able to reduce population growth. We must accept the differences among us and learn to set appropriate boundaries--without violating or destroying each other. Although fear and some degree of danger may be involved, we cannot afford for fear to control our behavior, for this will destroy our integrity and our ability to be rational.

We must develop respect for the rights of others, even when we may not understand every aspect of a problem. Our emotional intelligence, rational intelligence and higher consciousness need to integrate in order for us to make choices with the least likelihood of causing negative consequences. Of prime importance is that we all work together to solve problems, from putting out fires under a burning bus to safely dismantling nuclear weapons.

Our commitment to sharing with compassion whatever resources we possess needs to take priority over our personal or national ambition for power and control over others. In the final analysis, "mi casa es su casa" is the ultimate truth. Our great work is to assure that our planetary home will be sustainable and peaceful for those who follow us.

While human behavior is influenced strongly by evolved traits, each of us is also influenced by how we perceive and interpret any set of circumstances. If we believe that we can continue using the Earth to meet our needs without considering hers, we will participate in destroying ourselves as we destroy our planet. If we believe that wars remain inevitable, and therefore refuse to commit to the sacrifices required to change our ways, the chances are high that a nuclear war will annihilate us and our living planet simultaneously. We each must adhere to the mind-set that a world without war is a possibility—one that we are capable of achieving.

We, Dolf and Paddy, urge you, the reader, to take on your corner of the world in your own individual way to work seriously to help create a sustainable planet in a world free of war.

We each have to be the change we wish to see in the world.
~ Gandhi

References:

Q. In *Power Quotes*. Baker, Daniel B. (Editor) USA: Visible Ink Press for Barnes and Noble Books, 1992.

1. Berry, Thomas. *The Great Work: Our Way into the Future*. New York: Bell Tower of Random House. 1999.

2. Pearce, J.C. *Evolution's End: Claiming the Potential of our Intelligence*. San Francisco: Harper, 1992.

3. Keen, Sam, *Faces of the Enemy*. New York: Harper and Row, 1988.

4. Hauser, M. D. *Moral Minds*. New York: Harper Collins, 2006.

5. Zinn, Howard, *The Politics of History, 2nd Edition*. Urbana and Chicago, Illinois: University of Illinois Press, 1990.

6. Mortenson, Greg, and Relin, David, *Three Cups of Tea*. New York: Penguin Books, 2006.

7. Hedges, Chris. *War is a Force that Gives Us Meaning*. New York: Public Affairs Press of Perseus Books Group, 2002.

8. Drakulic, Slavenka, *They Would Never Hurt a Fly*. Great Britain: Abacus Books, 2004.

9. Frankl, Victor, *Man's Search for Meaning*. New York: Simon and Schuster, 1984.

10. Beah, Ishmael, *A Long Way Gone: Memoirs of a Boy Soldier*. New York: Sarah Crichton Books, 2007.

11. Furuichi, Takeshi, In The New Yorker magazine, July 30, 2007.

12. de Waal, Frans. *Bonobo: The Forgotten Ape*. Berkeley and Los Angeles, California: University of California Press, 1997.

13. Ousby, I. (Editor) The Cambridge Guide to Literature in English. *Everyman*. Cambridge, England: Cambridge University Press, (2006)

14. Hegel, G.W.F. In Keen, Sam, *Faces of the Enemy*. New York: Harper and Row, 1988.

15. LeShan, Lawrence, *The Psychology of War*. New York: Helios Press, 2002.

16. Gorbachev, Mikhail. Greensboro, North Carolina: Bryan Lecture Series of Guilford College, October 6, 2004.

Acknowledgements

Rudolf Harmsen

When I retired from my academic position, I decided to refocus my life towards writing for the educated general public about the scientific topics closest to my heart. Over thirty-five years of developing expertise in evolutionary biology I decided to write more than articles in scientific journals. While working as a lecturer on one of Lindblad's expedition ships, honing my skills for introducing people other than students to the science behind natural history, I met Paddy. Soon our discussions focused on the interaction of psychology and evolution, and the seeds for *Love and War: Human Nature in Crisis* were sown. Co-authoring a book that straddles two disciplines is a learning experience for both writers, and learning we did. It also leads to inevitable arguments; but when people argue with resolution as their goal, the end result is much more than the sum of the original opinions. I believe that we have ended all of our arguments in this manner, and I thank Paddy for being the most argumentative co-author, but also the best I could have wished for.

One always learns from one's colleagues, and I was an inveterate picker up of learner's crumbs. The colleague and friend who influenced my understanding of evolutionary biology most is George Williams. Over several years, he and I have walked countless miles talking about the intricacies of the various aspects of natural selection and how all this has shaped human nature. Thank you George for this invaluable gift.

More recently, I have benefitted enormously from the comments and criticism of drafts of the book provided by experts in a broad spectrum of disciplines, from biologists and psychologists to specialists in peace studies and ministers of religion. I am especially grateful for the intelligent criticism of Bill Roff, who made me rethink some major concepts, improving the book significantly.

Finally, I thank my wife Jeri for her patience and her love. The book meant that we had to rethink the first few years of our retirement, as I spent many hours alone with my word processor, devoting perhaps too much passion on writing, and not enough on our relationship.

Paddy S. Welles

This book could never have happened without my having met Dolf on a Lindblad cruise, so I remain grateful to Lindblad Expeditions for planning the Alaska Cruise in June 2002, and for having him as a member of their teaching staff. My second book on love-relationships had recently been published, and I was searching for a way to give this topic more latitude. Simultaneously, Dolf expressed a desire to write a book for publication that could help end war. Because the world was still reeling from the effects of 9/11, we decided to combine our fields of expertise, psychology and biology, to create an in depth understanding of why our species still continued to believe that violence and wars ever solved problems long term.

Since my publicist and agent, Kae and Jon Tienstra, without whose support and love I never would have made it through the publication of any of my books, encouraged me to delve into this topic, I began the arduous five-year task of co-authoring *Love and War: Human Nature in Crisis*. Co-authoring is filled with its own crisises, yet the process has stretched my mind, heart—and patience. I am deeply appreciative for all the lessons inherent in the experience and for the opportunity of working with Dolf, whose knowledge and perspective have been invaluable. Marian Sandmaier, our editor extraordinaire, has stuck with me from the beginning of my career as an author and has used up many red markers to help me (and us) improve my (and our) writing.

Author Sam Keen, remains my inspiration as a writer. He sets the bar beyond my reach, as he is simply the most brilliant and talented of all, and I am blessed by his friendship, tough criticism, and intermittent encouragement. Others who have influenced my thinking, added new ideas, and given encouragement throughout the writing of this book and to whom I owe bushels of appreciation are Harry Petrequin, Tim Welles (my husband, mostly for his ability to make my computer behave!), Peg Gallagher, Jerry Ackerman and Ellie Macklin, Carol Campbell, Charlotte Parker, Jan Rainier, Paul and Draha Krokosek, Stewart and Jarmila Peck, and Brent Sullivan, whose artistic son, Conor, created the rifle with flowers for the cover, which symbolically represents the message of the book.

A special kind of thanks goes to Dr. Patch Adams, who advised me to send this manuscript to Bob and Cleone Reed, Robert D. Reed Publishers, who said "yes" to sharing our dream of a more peaceful and sustainable planet. They, with typesetter, Amy Cole, are responsible for transforming our efforts into the book we hope will help end war and create a more compassionate world.

With love and gratitude to each of you, Paddy

Bibliography

1. Anderson, Hans Christian. *The Emperor's New Clothes*, adapted by Stephen Corrine in *Stories for Seven-Year-Olds*. London: Puffin Books, 1964.

2. Appiah, K.A., *The Ethics of Identity*. Princeton University Press, 2005.

3. Associated Press Release, Nov., 2006.

4. Atack, I. *The Ethics of Peace and War*. Edinburgh: Edinburgh Press. 2005.

5. Berreby, David. *Us and Them* In *The Science of Identity*. Chicago: University of Chicago Press, 2008.

6. Baker, Daniel B. (Editor) *Power Quotes*. USA: Visible Ink Press for Barnes and Noble Books, 1992.

7. Bartlett, John. *Familiar Quotations, 13th Edition*. New York: Little, Brown and Company, 1955.

8.	Boehm, Christopher. *Hierarchy in the Forest*. Cambridge, Massachusetts: Harvard University Press, 1999.

9.	Bloom, Howard. *The Lucifer Principal*. New York: The Atlantic Monthly Press, 1995.

10.	Bowlby, John. *Attachment and Loss*. New York: Basic Books, 1969.

11.	Bradley, James. *Flags of Our Fathers*. New York: Random House. 2000.

12.	Bradly, James. *Flyboys: A True Story of Courage*. New York: Little, Brown and Company, 2003.

13.	Brown, Lester. *Plan 3.0: Mobilizing to Save Civilization*. New York: W. W. Norton & Co, 2008.

14.	Bugos, Paul E. and McCarthy, Lorraine M. *Ayoreo Infanticide: A Case Study*. In Hausfater, Glenn and Hrdy, Sarah (Editors) *Infanticide: Comparative and Evolutionary Perspectives*. New York: Aldine and Gruyter Press, 1984.

15.	Byrd, Robert, U. S. Senate Floor remarks, May, 2003.

16.	Carter, James E., *Our Endangered Values*. New York: Simon and Schuster, 2005.

17.	Chagnon, Napoleon. *Yanomamö: The Fierce People*. Fort Worth, Texas: Harcourt, Brace Publishers, Inc., 1997.

18.	*Civil Paths to Peace*. Report of the Commonwealth Commission on Respect and Understanding. Commonwealth Secretariat, 2007.

19.	Clemenceau, George. In Baker, Daniel B. (Editor) *Power Quotes*. USA: Visible Ink Press for Barnes and Noble Books, 1992.

20.	Daly, Martin and Wilson, Margo. *Homicide*. New York: Aldine de Gruyter Press, 1988.

21.	Dawkins, Richard. *The God Delusion*. New York: Houghton-Mifflin. 2006.

22. Demoulin, Stephanie, et al. *Intergroup Misunderstanding: Impact of Divergent Social Realities.* London: Psychology Press, 2008.

23. de Mello, Anthony. *Awareness: The Perils and Opportunities of Reality.* New York: Doubleday, 1990.

24. de Waal, Frans. *The Age of Empathy: Nature's Lessons for a Kinder Society.* New York: Random House, 2009.

25. de Waal, Frans. *Good Natured: The Origins of Right and Wrong in Humans and Other Animals.* Cambridge, Massachusetts: Harvard University Press, 1996.

26. de Waal, Frans. *Our Inner Ape.* New York: The Berkeley Publishing Group, 2006.

27. de Waal, Frans. *Primates and Philosophers: How Morality Evolved.* Princeton University Press, 2006.

28. Deffeyes, K. S. *Beyond Oil—The View from Hubbert's Peak.* New York: Hill and Wong, 2005.

29. Donne, John. In Bartlett, John, *Familiar Quotations, 13th Edition.* New York: Little, Brown, and Company, 1955.

30. Durant, William. In Wall, Ronald. *Sermons for the Holidays.* Grand Rapids, Michigan: Baker Book House, 1989.

31. Dyer, Gwynne. *War.* New York: Crown Publishers, 1985. PBS National Television Series. 1986.

32. Evans, Paul and Bartolome, Fernando. *Must Success Cost so Much?* London: Grant & McIntyre, 1980.

33. Fisher, Helen. *The Anatomy of Love.* New York: W. W. Norton, 1992.

34. Fisman, R. and Miguel, E. *Economic Gangsters.* New Jersey: Princeton University Press. 2009.

35. Fossey, Dian. *Infanticide in Mountain Gorillas*. In Hausfater and Hrdy, (Editors) *Infanticide: Comparative and Evolutionary Perspectives*. New York: Aldine de Gruyter Press, 1984.

36. Frank, Robert H. *The Strategic Role of the Emotions*. New York: W. W. Norton. 1988.

37. Freud, Sigmund. *The Basic Writings of Sigmund Freud*. Translated and Edited by A. A. Brill. New York: Modern Library, 1938.

38. Fry, Douglas P. *The Human Potential for Peace: An Anthropological Challenge to Assumptions about War and Violence*. Oxford: Oxford University Press, 2006.

39. Gat, Azar. *War in Human Civilization*. Oxford: Oxford Universtiy Press, 2006.

40. Global Issues, Jan. 2008.

41. Golding, W. G. *Lord of the Flies*. United Kingdom: Faber and Faber, 1954.

42. Goleman, Daniel. *Emotional Intelligence*. New York: Bantam Books, 1995.

43. Goodall, Jane, with Berman, Phillip. *Reason for Hope*. New York: Warner Books, 1999.

44. Gould, Roger. *Transformations*. New York: Simon & Schuster, 1978.

45. Gray, John. *Men are from Mars and Women from Venus*. New York: Harper-Collins,1992.

46. Gregor, T. *Uneasy Peace: Intertribal Relations in Brazil's Upper Xingu*. In Haas, J. (Ed) *The Anthropology of War*, Cambridge University Press. 1990.

47. Harlowe, Harry F. "Basic Social Capacities of Primates," in *Human Biology*, 1959, Vol. 31

48. Harvey, Anthony, *The Lion in Winter*. Study Guide, BookRags. The Gale Group, Inc. Farmington, MI, 2000.

49. Hauser, Marc. *Moral Minds: How Nature Designed Our Universal Sense of Right and Wrong*. New York: Ecco Press, 2006.

50. Hausfater, Glenn and Hrdy, Sarah (Editors) *Infanticide: Comparative and Evolutionary Perspectives*. New York: Aldine de Gruyter Press, 1984.

51. Hedges, Chris. *War is a Force that Gives Us Meaning*. New York: Public Affairs Press of Perseus Books Group, 2002.

52. Hoffman, Robert. *Everyone is Guilty, But No One is to Blame*. Oakland, California: Recycling Books, 1988.

53. Holy Bible.

54. Internet report. info@votevets.org. 2002.

55. Jansen, M. (Ed.) *Warrior Rule in Japan*. Cambridge, England: Cambridge University Press, 1995.

56. Keeley, L. H. *War before Civilization, The Myth of the Peaceful Savage*. Oxford University Press. 1996.

57. Keen, Sam. *Faces of the Enemy: Reflections of the Hostile Imagination*. First published New York: Harper and Row, 1988. Revised, expanded Edition, 2004.

58. Keen, Sam. Lecture on *The Passionate Stages of Life*, Washington, D.C., 1999.

59. Keen, Sam. *The Passionate Life*. New York: Harper Collins, 1984.

60. Kelley, Thomas P., *The Black Donnelly's*. Firefly Books, Buffalo, 1993.

61. Kenneally, Christine. *First Word*. New York: Penguin, 2007.

62. Kline, T., Edwards B., and Wymer, T., *Searching for Great Ideas*. New York: Harcourt, Inc., 1998.

63. Kimbrell, A. (Ed.) *Fatal Harvest—The Tragedy of Industrial Agriculture.* Washington, D.C., Island Press. 2002.

64. Knowles, John. *A Separate Peace.* New York: Macmillan, 1959.

65. Kopp, Sheldon. Eschatological Laundry List in *Hidden Meanings.* New York: Science and Behavioral Books, Inc. 1976.

66. Lawrence, T. E. *Lawrence of Arabia.* In *Seven Pillars of Wisdom.* Library Edition. Castle Hill Press, 1922.

67. Le Blanc, S. A. *Constant Battles, Why We Fight.* St. Martin's Griffin. 2003.

68. Lee, Robert E. In Bloom, Howard. *The Lucifer Principal.*

69. Lewis, Bazett. *The Murle Red Chiefs and Black Commoners.* Oxford, UK: Clarendon Press, 1972.

70. Lewis, C.S. *A Grief Observed.* New York: Harper Collins, 1961.

71. Lozoff, Bo. *We're All Doing Time.* Durham, North Carolina: Human Kindness 1985.

72. MacGowan, Christopher. *The Collected Works of William Carlos Williams.* New York: New Directions Publishing Co., 1986.

73. Madison, James, in Political Observations, 1795.

74. Meadows, D. L. and Meadows, D. H. *The Limits to Growth— Club of Rome Project on the Predicament of Mankind.* New York: Universe Books, 1972.

75. Mech, L. D. *The Arctic Wolf—Ten Years with the Pack.* Stillwater, Minnesota: Voyageur Press, 1997.

76. Mech, L. D. *The Wolf—The Ecology and Behavior of an Endangered Species.* Minneapolis, Minnesota: University of Minnesota Press, 1981.

77. Meggitt, M.J. *Desert People, A Study of the Walbiri Aborigines of Central Australia.* Un. Of Chicago Press. 1965.

78. Moore, Thomas. *Soul Mates.* New York: Harper-Collins, 1994.

79. Muir, Edward. *Mad Blood Stirring: Vendetta and Factions in Friuli during the Renaissance.* John Hopkins University Press, 1993.

80. N.Y. Times Magazine

81. Obama, Barack. *The Audacity of Hope: Thoughts on Reclaiming the American Dream.* New York: Random House Audio, 2006.

82. Orwell, George. *Animal Farm.* London: Secker and Warburg, 1945.

83. Osler, Sir William. *Science and Immortality.* In Bartlett, John, *Familiar Quotations, 13th Edition.* New York: Little, Brown, and Company, 1955.

84. Otterbein, K.F. *How War Began.* College Station, Texas: A&M Univ. Press, 2004.

85. Ousby, Ian, *Cambridge Guide to Stories in English.* England: Cambridge University Press, 1983

86. Owen, Wilfred. Preface to Collected Works. New York: New Directions Book, 1963.

87. Paine, Thomas. In Baker, Daniel B. (Editor) *Power Quotes.* USA: Visible Ink Press for Barnes and Noble Books, 1992.

88. Pearse, F. The Bog Barons—Scandal in the Jungle. *New Scientist* 196: 50, 2007.

89. Petrequin, Harry. National War College papers.

90. Popper, Karl, *The Self and Its Brain.* New York: Springer Publishing, 1985.

91. Popper, Sir Karl, *The Open Society and Its Enemies.* Fifth Edition. Routledge & Kegan, 1966.

92. Psych. Today (and medical encyclopedia)

93. Pusey, Anne E. *Of Genes and Apes: Chimpanzee Social Organization and Reproduction*. In de Waal, F. B. M. (Ed)—2001—*Tree of Origin—What Primate Behavior Can Tell Us about Human Social Evolution*. Cambridge, Mass.: Harvard University Press, 2001.

94. RAF Bomber Command Diary. In Bradley, James, *Flyboys*.

95. Rees, M. *Our Final Hour: A Scientist's Warning: How terror, error, and environmental disaster threaten humankind's future in this century—on Earth and beyond*. New York: Basic Books, 2003.

96. Renfrew, Colin, *Prehistory: The Making of the Human Mind*. New York: Modern Library, 2007.

97. *Report of the Canberra Commission on the Elimination of Nuclear Weapons*. 1997. http://prop.org/2000/canbrp01.htm

98. Rosny, J. H. *La Guerre du Feu*. In Wright, Richard, *A Short History of Progress*.

99. Sagan, Carl. *Billions and Billions*. New York: Ballantine Books, 1998.

100. Sandlin, Jo (Ed.), *Bonobos: Encounters in Empathy*. Milwaukee: Zoological Society of Milwaukee, 2007.

101. Sevile, George (Lord Halifax) in Bartlett, John, *Familiar Quotations, 13th Edition*. New York: Little, Brown, and Company, 1955.

102. Shakespeare, William. (from *Hamlet* and *Romeo and Juliet*) in Bartlett, John, *Familiar Quotations, 13th Edition*. New York: Little, Brown, and Company, 1955.

103. Shakespeare, William. "Measure for Measure," in *The Works of William Shakespeare, Complete*. Ed. by Walter J. Black. New York: Black's Readers Service, 1944.

104. Sheeran, Josette, in UN World Report, 2008.

105. Sherif, M. et al. 1988. *The Robbers Cave Experiment, Intergroup Conflict and Cooperation*. Wesleyan U. P.

106. Shermer, Michael. *Why Darwin Matters*. New York: Holt McDougal, 2007.

107. Siegel, Daniel. *The Developing Mind*. New York: Guilford Publications, 1999.

108. Sinnot-Armstrong, Walter, (Ed.) Moral Psychology, Vol. 1: *The Evolution of Morality: Adaptations and Innateness*. Cambridge, MA: MIT Press. 2007.

109. Smith, Adam. *The Theory of Moral Sentiments*. In Oxford University Magazine, 1759.

110. Tennov, Dorothy. *Love and Limerance*. New York: Scarborough House, 1999.

111. Texas Social Services Report, 1997.

112. Thomson, Oliver. *The Great Feud—the Campbell's and MacDonald's*. Sutton Publishing, 2001.

113. Thurman, Howard. In Keen, Sam. *The Passionate Life*. New York: Harper Collins,1984

114. Thurman, Robert. *Inner Revolution*. New York: Riverhead Books, 1998.

115. Tobin, James. *Ernie Pyle's War*. New York: Free Press, 1997.

116. Tolstoy, Leo. *War and Peace*. Translated by Alexandra Kropotkin. Great Britain: The John C. Winston Company, 1949.

117. Walker, Gabrielle and King, David. *The Hot Topic: What Can We Do About Global Warming and Still Keep the Lights On?* London: Bloomsbury/Harcourt, 2008.

118. Warner, W. L. *A Black Civilization. A Study of an Australian Tribe*. Harper and Row. 1958.

119. Watkins, Trevor. *The Beginnings of Warfare*. In Hackett, J. (Ed.) *Warfare in the Ancient World*. New York: Facts on File Publishing, 1998.

120. Welles, Paddy S. *Are You Ready for Lasting Love?* New York: Marlowe and Company. 2002.

121. Wells, H. G. in Bartlett, John, *Familiar Quotations, 13ᵗʰ Edition.* New York: Little, Brown, and Company, 1955.

122. Welzer. M. *Just and Unjust Wars: A Moral Argument with Historical Illustrations.* 2nd Edition. New York: Harper Collins. 1992.

123. Wendorf, Fred. *The Prehistory of Nubia.* Vol. 2. Dallas, Texas: Southern Methodist University Press, 1968.

124. White, Ralph, Editor. *Psychology and the Prevention of Nuclear War.* 1986.

125. Wilson, E. O. *The Future of Life.* New York: Knopf, 2002.

126. Woodwell, G. M. (Ed.) *Forests in a Full World.* New Haven: Yale University Press, 2001.

127. Wrangham, Richard. "The evolution of sexuality in chimpanzees and bonobos." *Human Nature,* 4(1), 47–79. 1993.

128. Wright, Ronald. *A Short History of Progress.* Toronto, Canada: House of Anansi Press, 2004.

About the Authors

Rudolf Harmsen, Ph.D.

Author Photographer: Keith Skelton

I was born in the Netherlands in the year Adolf Hitler gained power in Germany. My childhood was spent in Nazi-occupied Holland, and I experienced acts of violence most children can't even imagine. As a nineteen-year old, I was drafted into the Dutch army and served with NATO forces in Germany, being trained to fight the Soviets in case the cold war turned hot.

After a brief period at a Dutch university, I immigrated to Canada and completed a BA and an MA in Biological Sciences at the University of Toronto. Three years of doctoral studies in insect biochemistry in Cambridge (UK) gained me a Ph.D., which was followed by a post-doc at the University of East Africa in Nairobi, Kenya. It was there that I found the intellectual stimulation I needed to mature into a creative scientist by hobnobbing with a crowd of expatriate anthropologists such as Louis Leaky and Glynn Isaac, and primatologists such as Stewart Altman, Thelma Rowell, and Jane Goodall.

Post-docs, unfortunately don't last forever, and by 1966 I left Africa, and joined the faculty at Queen's University in Kingston, Canada. Here I completed my transformation to an evolutionary ecologist by working with a group of young, very enthusiastic colleagues, and by maintaining close contact with some intellectual giants. I especially learned a great deal from men such as George Williams and Larry Slobodkin, both of Stony Brook. Sabbatical leaves in Vancouver to learn computer simulation methods, Australia and New Guinea to delve into tropical rainforest ecology, and Oxford to once again hone my theoretical evolutionary discussion skills, allowed me to constantly update my knowledge and understanding of the biological sciences.

My own research was concept oriented, which means that I was part armchair biologist, creating new theories, and part experimentalist. Jointly with my graduate students, we tested our theories with a wide variety of organisms, all the way from moose and muskox in the arctic, to spider mites and leaf miners in the laboratory. Throughout my career as a scientist, however, I kept close contact with colleagues in anthropology, psychology, and philosophy, developing a broad understanding of issues in which evolutionary biology and human behavior interact. This meant that I am living a life committed to the understanding of the human condition: the evolution and cultural development of the most interesting life form on Earth. So far, other than over a hundred scientific publications, I have written mostly for myself. Once in a while I have dropped everything and written a letter-to-the-editor of some newspaper, which each time got me into difficulties as many readers are deeply offended by my unorthodox opinions. Now, retired from my university position, I organize ecology tours to out-of-the-way places, and have started to write my ideas on the human condition for possible publication.

In 1960 I married Jeri Shortt, who became an art museum educator of note, and we have two sons, one is a dramatic artist in New York, the other a visual artist in Toronto. I also have a different kind of offspring, my ex-grad students, who fill professorships and other positions in the world of science in several countries, including the US, the UK , Canada, Australia, and New Zealand .

My wife and I live on a reforested hundred-acre farm in Glenburnie, Ontario, just north of Upstate New York, where I write, garden, and spend too much time fixing our 185-year-old farmhouse.

You may write to me at harmsenr@queensu.ca.

Paddy S. Welles

Author Photographer: Personius-Warne

My earliest memories center around the air raid drills of WWII when our family would huddle under a tent constructed of blankets on the tiny back porch of our home in Sanford, NC. My father would hold a small, lighted candle and try to explain "war." My immature mind just could not grasp why people would kill each other just because they couldn't agree on something. Today my more mature mind is still haunted by the same question, especially when I catch myself in anger thinking that I wish the source of my anger (usually a person I love, or they would not have had the power to hurt me) would just drop dead. At some early age, these concepts of love and war began to weave a mysterious tapestry that has hung in the shadows of my thinking for years.

I pursued degrees in Psychology, Sociology (UNC, Chapel Hill), and a Ph.D. in Child and Family Studies (Syracuse Un.), have taught at almost every level of the American educational system, from first grade through graduate school, and have been a Marriage and Family Therapist in private practice for many years in upstate NY. After listening carefully to thousands of individuals' stories, to both the tragic and wonderful things that happen in families and in communities, to the rhythms of my own heart and soul, I have begun to untangle that mysterious tapestry.

The dominant thread is that all life is truly one—and a more miraculous one than we can possibly imagine.

My four children born during my first marriage of fifteen years, two stepchildren acquired with my second marriage of over thirty years, and eight grandchildren have taught me more about love and war than I ever wanted to know. Simultaneously, living in France for almost four years in the late 1970s expanded that knowledge to validate that regardless of whom we are, whatever our life circumstances, wherever we live, we share our deepest psychological need of wanting our existence acknowledged. This need is most fully satisfied when we feel loved and are able to respond to love. There seems to be a shared universal desire in each of us to make a mark that verifies we do indeed exist. My desire to this end has compelled me to write about love and relationships in two previously published books, *To Stand in Love* and *Are You Ready for Lasting Love?* (This book is now also published in Portuguese and is listed as a best seller on the Internet). However, I believe this book, *Love and War*, co-authored with Dolf Harmsen, will be the mark that *can* make a lasting difference—to bring us closer to real love and lessen the chances of our destroying the civilized world.

Please visit me at my blog: http://paddy-loveandwar.blogspot.com or write to me at paddysw@aol.com.

Robert D. Reed Publishers Order Form

Call in your order for fast service and quantity discounts!
(541) 347- 9882

OR order on-line at **www.rdrpublishers.com** *using PayPal.*
OR order by FAX at **(541) 347-9883** *OR by mail:*
Make a copy of this form; enclose payment information:
Robert D. Reed Publishers
1380 Face Rock Drive, Bandon, OR 97411

Send indicated books to:

Name: _____

Address: _____

City: _____ State: _____ Zip: _____

Phone: _____

Fax: _____

Cell: _____

E-Mail: _____

Payment by check /_/ or credit card /_/ (All major credit cards are accepted.)

Name on card: _____

Card Number: _____

Exp. Date: _____ Last 3-Digit number on back of card: _____

	Quantity	Total Amount
Love and War by Rudolf Harmsen and Paddy Welles ... $17.95	_____	_____
Fearless by Steve Chandler $12.95	_____	_____
Shift Your Mind: Shift the World by Steve Chandler ... $14.95	_____	_____
The New American Prosperity by Darby Checketts ... $12.95	_____	_____
The Secret Sin of Opi (novel) by Peter Cimini $24.95	_____	_____
The Media Savvy Leader by David Henderson $19.95	_____	_____
How Bad Do You Really Want It? by Tom Massey $19.95	_____	_____

Quantity of books ordered: _____ Total amount for books: _____

Shipping is $3.50 1st book + $1 for each additional book: Plus postage: _____

FINAL TOTAL: _____